Your Comprehensive Guide to Managing an IT Department

Dr. Julia Diakakis & Saki Diakakis

ISBN: 978-1-962557-03-0

DEDICATION

This book is dedicated to all our friends and family for their support and tolerance over the years. Special thank you to Scott Barber who unknowingly provided inspiration and insight along the way.

Table of Contents

Introduction

This book is designed for those wishing to create and maintain a successful Information Technology (IT) department. Whether you are a business owner looking to upgrade your infrastructure or an information technology professional wanting to better understand the fundamentals of effective IT management, this guide provides the essential steps needed for building and sustaining a strong IT team. Included are topics ranging from the establishment of clear goals and objectives to the development of an effective leadership plan, including the utilization of automation as needed, implementation of continuous education programs, and regular monitoring of performances. With detailed explanations of each step and actionable advice, this book will provide readers with valuable insights into making IT departments run smoothly and efficiently. Reading this book will provide you with comprehensive and proven strategies for maintaining a successful IT department.

These key strategies have been tested and proved in the business world with great success - this book will give you access to these same techniques so that you can apply them to your organization. In addition, this book provides detailed explanations of each step along with actionable advice to ensure readers are equipped with the knowledge needed to make IT departments function smoothly and successfully.

IT Hurdles

Introduction

Many IT departments face a variety of challenges daily. From having outdated technologies to dealing with security threats, IT departments often find themselves struggling to keep up with the ever-evolving landscape of technology. One of the biggest challenges is staying up to date with the latest tools needed to maintain high levels of productivity. Additionally, IT departments must be mindful of budgeting as sufficient resources are needed to stay ahead in the field. Security is another major concern for many IT departments, as they must be vigilant in protecting their systems and data from malicious actors. Finally, optimizing existing software and hardware while aligning with company goals can be quite demanding for IT professionals.

Outdated Technologies

Outdated technologies can have a huge impact on an IT department or the entire company. Outdated software or hardware can cause delays, malfunctions, decreased productivity, and even security vulnerabilities that put sensitive company data at risk. To keep up with technological advances, IT departments must invest in new tools and technologies, and sometimes training.

Staying current often requires dealing with budget constraints as the cost of new systems can be problematic for already strained budgets. However, maintaining outdated systems can also be costly due to their lower efficiency levels and the need for additional maintenance and technical support over time.

When organizations opt to retain outdated technology instead of investing in more modern alternatives, individuals in IT roles need to understand how this decision might affect the company's overall performance and potential for future business opportunities. Outdated technology may limit collaboration efforts between teams or hinder customer service standards if customers are unable to receive timely responses from organizations relying on old systems. Legacy systems may also lead to missed deadlines due to poor performance or system compatibility issues thus requiring further troubleshooting from IT professionals.

Outdated systems may also pose significant cybersecurity threats leaving sensitive data exposed and vulnerable to malicious actors who leverage out-of-date software designs that haven't been updated with current security protocols. Failing to invest in modern solutions, when necessary, puts organizations at risk of being victimized by cyber-attacks seeking weak points in older IT infrastructures.

Overall, organizations and IT departments need to understand how retaining older technologies could negatively impact operations both now and in the future. Investing in modern tools that match industry standards and technological advancements allows companies to better protect their businesses while staying competitive.

Keeping Up with the Latest Tools and Technologies

To stay competitive IT departments are invariably tasked with finding ways to keep up with the latest tools and technologies. Yet, staying informed of the most recent trends can be challenging. With so many new developments in the tech space, it is nearly impossible for individuals in an IT role to manually stay up to date on every piece of software and hardware that could be benefit their organization.

Additionally, as previously mentioned, budget constraints are often a factor when it comes to investing in the newest technology tools. Organizations may opt for less expensive, inferior alternatives or even hold on to outdated systems if it appears that the cost of upgrading is not equitable to the potential returns. Unfortunately, this can hinder business growth as new advancements are not optimized for outdated systems and therefore limit collaboration efforts between teams or preclude customer service standards.

To remain competitive and leverage technological advantages within their industries, organizations must find ways to keep up with current trends without breaking budgets. This includes researching existing tools, exploring multiple options before selecting a solution, and utilizing cost-effective strategies such as cloud hosting or subscription services whenever possible. Moreover, IT departments should expose members of their team to any new releases or advancements in cybersecurity protocols so they can best advise higher-level executives before making any updates or investments—allowing them to make more informed decisions that do not compromise security or business objectives.

Overall, keeping up with the latest tools and technologies is essential yet challenging for IT departments wanting to ensure their organizations remain competitive. By dedicating resources to industry research and staying abreast of new developments whenever possible, IT professionals can equip themselves with up-to-date knowledge while equipping their businesses with the necessary solutions needed for long-term success.

Security Threats

Security threats are an ongoing struggle for IT departments. As technology continues to evolve, so do the associated risks. Unfortunately, many businesses often don't realize that their existing systems may be vulnerable until a breach has already occurred. This can lead to costly damages as well as lost customer trust and loyalty which ultimately affects the reputation of a company.

To ward off potential security threats and protect corporate or customer data, IT departments must invest in effective cybersecurity solutions such as robust anti-virus software, firewalls, and encryption tools. However, this requires significant resources—which may not always be available due to budgetary constraints. Moreover, companies must stay up to date on the latest security protocols to best protect themselves against emerging threats such as phishing attempts or malware infections.

Overall, organizations should prioritize their security measures if they want to protect their data and maintain customer trust. By researching industry trends and investing in cost-effective solutions whenever possible, IT professionals can help organizations stay ahead when it comes to safeguarding sensitive information while also working within budget.

Optimizing Existing Software and Hardware

Optimizing existing software and hardware is yet another challenge faced by IT departments. With advancements in technology, older systems may not be able to compete with the speed or features of newer versions—resulting in potential issues with collaborative efforts between teams or delayed customer service standards.

To maximize the potential efficiency of existing systems, organizations must research and integrate cost-effective solutions that do not require extensive investments or hardware modifications. This may include cloud hosting services that allow users access to greater processing power without additional outlay, subscription plans which offer discounts on multiple licenses, and open-source resources that provide free software upgrades.

IT workers should investigate any available updates for existing systems to ensure they are using all the most recent security protocols and performance optimizers. Organizations may also need to increase memory or storage capacity if these capabilities are limited by their current system—which requires a holistic view of their existing infrastructure before making any purchases.

Overall, while it can be daunting to optimize older systems without causing any major disruption to operations, IT departments must continue researching industry trends as well as exploring multiple options that reduce cost before deciding on an effective solution for business needs. Dedicating resources to finding compatible tools and implementing strategies that extend the life of existing software and hardware, allows IT professionals to help organizations remain competitive while ensuring long-term success.

Develop a Plan

The Business Plan

Developing a business plan for IT is an important step for any IT department striving for success. A well-developed business plan provides the department with a clear understanding of the goals and objectives, how to achieve these goals and objectives, and how to accurately measure success. An effective IT business plan takes into account all aspects of an organization's technology needs and provides a road map for necessary actions and future decisions.

A well-crafted IT business plan guides an organization in several ways:

- Increased efficiency through the identification of areas where technology can streamline processes, such as data entry or reducing paperwork.
- Creation of better customer service thereby improving process flows and providing customers with faster access to services.
- Increased productivity thereby leveraging technology to automate mundane tasks, thus freeing time for team members' time to attend to valuable activities.
- Saved money through the optimization of hardware purchasing and upgrading specifications to receive the most from current resources without overspending on new technology.
- Identification of potential opportunities for growth thereby setting goals and developing strategies for achieving them.

• Proactively managing risk through the anticipation of future needs and creation of plans to address potential threats before they become problems.

An effective IT business plan should provide detailed information on the organization's current state, desired outcomes, short-term strategies, long-term goals, risk management strategies, budget needs, cost estimations, and time line projections. All stakeholders should be included in the planning process as a method of empowerment and ownership to further impart the importance and awareness of developing a comprehensive plan. This furthers buy in and stake when defining concise goals and objectives to achieve those goals. With a solid IT business plan in place, organizations ensure that their technology investments are strategically sound with minimal disruption or downtime due to technical issues.

How to Develop an IT Plan

Developing an IT business plan is a multi-step process. Listed below are suggested steps to take when creating a comprehensive IT business plan:

1. Define goals and objectives: Identify and list the organization's goals and objectives. These should include short-term goals such as increasing efficiency or customer service, as well as long-term goals such as growth and expansion.

2. Assess the environment: Gather data on the current technology environment of the organization, including hardware inventory, software used, network architecture, and security practices. This determines what needs to be upgraded and where resources can be optimized for the best results.

3. Brainstorm solutions: Create a list of potential solutions that can be used to achieve the identified

objectives. Be sure to consider budget constraints and vendor relationships to make realistic decisions about potential solutions.

4. Create action items: Create a timeline with specific tasks that need to be completed to achieve each goal outlined in the plan. Assign responsibilities for each task and develop milestones for tracking progress throughout the implementation of the business plan. This will hold stakeholders accountable and ensure timelines are met.

5. Establish metrics: Set benchmarks for measuring success, both short-term (measured quarterly) and long-term (measured annually). This will provide hard evidence of whether or not strategies outlined in the business plan were effective in reaching the desired outcomes or if adjustments need to be made.

6. Monitor performance: Regularly review performance metrics against established benchmarks to assess whether changes are needed or if strategies have been successful for reaching goals outlined in the business plan. This step is critical for assessing whether an organization's technology investments are paying off in improved efficiency, productivity, customer satisfaction, or other desired outcomes as stated in the original development of the plan.

A Coherent IT Department is Crucial

Having a coherent IT department is essential for any organization, regardless of the size. A coherent IT department allows for efficient use of resources, reduces duplication of efforts, and consistently works to improve communication among teams. A cohesive IT department helps create an environment where effective solutions can be identified and implemented quickly. Having a single point of contact for all IT-related questions also allows for faster resolution times and less confusion over responsibilities of personnel. Additionally, collaboration among departments within the IT team brings further benefits such as improved customer experience, better security practices, and enhanced productivity. By ensuring that the proper person is assigned to each task thus ensuring prompt completion, a cohesive IT department ensures that the organization's goals are met efficiently and effectively.

Having a cohesive IT department ensures consistency regarding approaches and best practices. Consistency ensures the organization is following a standardized set of business processes, reducing risk and increasing efficiency. Having a documented set of standard procedures, teams can quickly respond to changes in their environment or react to new threats. Additionally, personnel within the IT department can easily collaborate on projects, allowing for dynamic solutions.

With unified systems and processes in place, it becomes easier to identify inefficiencies and make improvements leading to better performance.

Ultimately, a cohesive IT department allows organizations to maximize resources while minimizing costly downtime or security risks. The benefits are not just limited to the IT team; improved communication among departments within an organization ensures benefits for customers with increased response times. Lastly, there is increased assurance that changes in strategy or operations will be implemented smoothly across the organization.

Policies, Procedures and Standards

Understand the Difference

Policies, procedures, and standards are important for the efficient and effective operations of any organization, and IT departments are no exception. Policies dictate the handling of different situations. Procedures explain how to execute something step by step. Standards ensure everyone is doing the same thing in the same way. Having these crucial and necessary pieces in place allows IT departments to operate effectively and efficiently.

Policies

Policies are an essential part of any organization, as they provide guidance and structure to ensure that the organization operates smoothly. Policies create clarity and consistency for the way things are done within the organization and aid in decision-making processes. When everyone is aware of the policies, they serve to reduce conflict, prevent misunderstandings and encourage efficient workflow.

Organizations without strong policies are more likely to experience issues with workflow and efficiency. There is an increased likelihood of communication breakdowns and misunderstandings among personnel, as well as possible confusion over job responsibilities and duties. Companies lacking clear policies tend to have disorganized operations and fewer opportunities for employee growth. Without proper guidelines in place, it is difficult for an organization to achieve success or even maintain a current level of performance.

In contrast, organizations with effective policies benefit from increased productivity due to improved organizational structure and clarity in how to conduct operations. With good policies in place, conflicts are minimized or avoided altogether, assignments are clearer, communication gaps are reduced or eliminated, and resources are used more efficiently and effectively. In addition, having clearly defined rules allows leaders to hold staff members accountable for their actions which creates a sense of responsibility among team members. Furthermore, well-defined policies encourage team unity as it fosters shared organizational values.

Overall, creating strong policies is crucial to operating and leading a successful organization to improved overall performance. Policies provide the necessary structure for greater efficiency and effectiveness while avoiding costly mistakes due to a lack of understanding of roles/responsibilities/duties among all personnel.

Examples of important policies for an IT department include:

- A data security policy outlining procedures for handling and storing sensitive information.
- An acceptable use policy establishing guidelines for using the computer network and other IT resources.
- A disaster recovery plan to ensure business continuity in the event of an emergency or natural disaster.
- A software license policy to prevent unauthorized use of purchased software.
- An access control policy outlining who has permission to access certain information.
- An email management policy to ensure proper use of company accounts.

• Policies regarding BYOD (Bring Your Own Device) usage on the company network.

Procedures

Procedures are essential, as they provide clear steps and guidelines that establish order and define the way things should be executed. Procedures help to ensure consistency in operations while helping reduce mistakes and misunderstandings that can arise from ambiguous instructions.

Organizations without strong procedures are likely to experience issues with workflow efficiency, miscommunication between employees, confusion over responsibilities and duties, and disputes over how tasks need to be completed. Companies lacking established processes tend to have fewer systems-oriented operations and fewer opportunities for employee development. Without consistent practices in place, it's difficult for an organization to establish itself as reliable or effective.

In contrast, organizations with effective procedures can benefit from improved productivity due to structured operations ensuring an awareness of the various roles within the company. With good procedures in place, conflicts are minimized or avoided altogether, assignments are easier to understand for all employees, communication gaps are reduced or eliminated, and resources are used more efficiently with less waste of time and/or money. In addition, having well-defined procedures allows leaders to hold their staff members accountable. Furthermore, having clearly defined procedures promotes team unity as they help create a shared understanding of steps to follow for any given policy implementation or routine task.

Overall creating strong policies is necessary for a successful organization and leads to improved operational successes. Procedures provide guidance that leads to greater clarity while helping avoid costly mistakes due to a lack of understanding of roles, responsibilities, and duties among its members.

Standards

Department and corporate standards are an important part of any organization, as they provide a framework to ensure that the company's operations, products, and services are of the highest quality. They help establish guidelines for employees to adhere to in producing consistent results and maintaining organizational control over aspects of the business. Standards create clarity and consistency in the way things are done within the organization and can aid decision-making processes set forth through company policies and procedures.

Organizations without clear standards may experience issues with workflow efficiency, customer dissatisfaction due to lower-quality products or services, confusion over job responsibilities and duties, and misunderstandings among departments or personnel. Companies lacking well-defined standards tend to have less stringent checks on quality assurance which can lead to costly mistakes. Without uniform criteria in place, it is difficult for an organization to ensure that its output meets industry standards or even maintains its current level of performance.

In contrast, organizations with effective standards benefit from increased productivity due to improved organizational structure and clarity in how operations should be conducted. With good standards in place, conflicts are minimized or avoided altogether, assignments are clearer for employees, communication gaps are reduced or eliminated, and resources are used more efficiently and effectively. Furthermore, having clearly defined standards allows leaders to hold staff accountable for their actions thus creating a culture of responsibility throughout the organization. Standards encourage everyone to take ownership of their work resulting in better quality output from everyone. Standards also promote unity as they foster shared values within an organization.

For example, if a customer service department has high standards such as having clear responses prepared beforehand and adhering strictly to timelines when responding to customer queries then it would not only guarantee customer satisfaction but also increase employee morale. In short, employees feel motivated when meeting these goals.

Overall creating robust standards is important to the operations of a successful organization and leads to improved operational success overall by ensuring high-quality work is produced consistently by every involved member. Standards provide structure leading to greater efficiency and effectiveness.

Summary

Policies, procedures, and standards are necessary elements for the success of any organization. Policies provide the structure for decision making throughout the company. Procedures provide step by step direction, allowing for a structured approach to operations that ensures consistency and minimizes mistakes due to misunderstandings or ambiguity. On the other hand, standards help organizations to maintain control over operations, products, and services while ensuring industry expectations are met. Organizations without strong policies, procedures and standards are likely to experience issues with productive workflow, efficiency, customer dissatisfaction, confusion over responsibilities and duties, and disputes over how tasks should be completed. With good policies, procedures and standards in place, conflicts are minimized or avoided altogether while resources can be used more efficiently and effectively with less waste. Finally, these practices promote team building through the creation of shared values and comprehensive guidelines as to how things should be done inside an organization.

Important Guidelines

Synopsis

Ensuring success for an IT department requires a combination of processes and systems, including:

• Clear goals and objectives with measurable outcomes, provide a basis to measure and assess progress against it. This can be done by setting clear Key Performance Indicators (KPIs) such as uptime targets, customer service resolution times, and response times.
• Regular check-ins within the team to identify any potential issues before they become major problems. This can include daily stand-up meetings, customer feedback surveys, and listening to customer complaints.
• Open communication among all members of the organization to ensure that any discrepancies in process or workflow are addressed quickly and effectively. Having a culture of collaboration and transparency ensures that everyone is on the same page with tasks and responsibilities.
• Automation systems of tasks such as bug tracking, issue reporting, code review, and project management. These tools streamline work flows and keep track of what everyone is working on at any given time.
• A reliable IT infrastructure with robust security measures in place to protect sensitive data. Setting up secure networks with regular system patches reduces the risk of cyberattacks or data breaches.

Getting Deeper into Goal vs Objective

Goals and objectives are two closely related concepts often used interchangeably in business. A goal is a broad, long-term aspiration that an organization sets out to achieve, while an objective is more specific and has an associated timeline. Goals provide the overall direction for an organization, while objectives act as steppingstones to reach those goals. Goals tend to be more abstract while objectives focus on measurable results. For an organization to achieve its goals, it must create objectives that will lead them in the right direction. Setting clear, achievable objectives with measurable outcomes assures everyone involved is doing their part to move towards those goals. Every successful business understands the importance of having both a set of well-defined objectives and the right strategies in place to reach those goals.

Setting objectives is an essential part of reaching any goal. Objectives provide a clear framework with measurable outcomes, which helps ensure that everyone involved is doing their part to reach the desired result. Without a well-defined set of objectives, it can be difficult to stay on track and measure progress toward the end goal. Objectives allow for better planning, as they give an organization a starting point and direction for the work needed to make the goal a reality. Additionally, setting objectives allows an organization to adjust its strategy if certain elements are not working as planned. In other words, objectives provide flexibility and help ensure that goals are ultimately achievable.

Plan to Reach Each Goal

- Set a goal: Define a clear and achievable goal for the organization.
- Create objectives: Identify measurable outcomes that will help reach that goal.

• Set timelines: Establish reasonable timelines to achieve each defined objective.
• Assign roles and responsibilities: Identify the resources required to hit each objective and create roles and responsibilities accordingly.
• Monitor progress: Track the progress of objectives to ensure they are in line with their intended timelines, strategies, and goals.
• Adjust strategy: If certain elements of the plan are not working as planned, adjust as necessary until the desired outcome is achieved or a new goal is set.

Identify Data Points to Track Progress

Data points to track progress towards an objective can include things like the number of tasks completed, percentage of tasks completed, amount of time spent on each task or project, and any other relevant information that measures the success of the overall goal. It is important to identify data points that are measurable and actionable - this will help you not only track progress but also improve in the future. Depending on your objectives, some data points may be more useful than others, so establish which data points are most important to monitor when setting your objectives and tracking progress.

Compare Data Points

Comparing data points can be an important and insightful tool to determine which are the most important to track toward reaching a goal. Analyzing and comparing data points, you pinpoint trends and identify areas of opportunity or weak spots. This will help you home in on the most relevant metrics for meeting objectives and help make informed decisions about how best to adjust your strategy to reach goals. Additionally, it is crucial to consider external factors such as market conditions or industry changes that may influence data points. Taking all this into consideration will give the best possible outcome when determining key data points.

Compiling relevant metrics for reaching goals is essential for maximizing chances of success. Start by identifying what objectives you want to reach, then research and analyze the data that is necessary to measure progress towards those objectives. Determine which metrics are most useful and have the strongest correlation with meeting those objectives - this may require comparing different data points. Once you have established key metrics, track them over time to gain insights into how you are progressing and adjust if needed. Consistently keeping an eye on these metrics will help you stay on track and ensure that any changes made are in line with the overall goal.

Adjusting based on data insights is key to reaching goals. Start by establishing clear objectives and identifying the relevant metrics that will best measure progress towards those objectives. Analyze any existing data to get a baseline understanding of how you are currently performing and identify trends or correlations. Track these metrics over time, looking for areas that could benefit from improvement or changes.

Once opportunities are identified, brainstorm potential solutions and determine which are the most viable ones in terms of cost, resources, and time. Brainstorming potential solutions is an important step toward reaching your goals. Start by thoroughly understanding the problem at hand and how it can be solved. Once a list is created, narrow down the list to potential solutions you believe will be most effective in achieving your objectives. Finally, execute the chosen solution and keep an eye on your progress to ensure it is having the desired effect.

Outline Timeline

Establishing a timeline for reaching any set of objectives is a fundamental part of any successful project. It enables the breakdown of large goals into smaller, more manageable steps, makes it easier to allocate resources and assign responsibilities, and helps keep everybody on track. A timeline keeps you from getting overwhelmed by setting realistic targets that are achievable in the given time frame. It also allows for flexibility when unexpected delays or events occur so your progress can be adjusted accordingly. To ensure success, create checkpoints throughout the timeline to review progress and make sure tasks are completed as planned. To summarize:

- Define the goal
- Break it down into smaller steps or objectives
- Identify resources needed
- Set due dates
- Create checkpoints throughout the time line
- Assign responsibilities
- Allow for flexibility
- Review progress regularly

Systems of a Successful IT Department

The Work Order System

A work order system, also known as a ticketing system, plays a pivotal role in the IT department's overall efficiency and effectiveness. It is an indispensable tool that streamlines several aspects of IT operations, from tracking issues and enhancing communication to coordinating resources and ensuring accountability. By centralizing the data repository and providing in-depth analysis and reporting, IT departments can continuously improve their performance and pinpoint areas of excellence.

One of the greatest benefits of implementing a work order system in an IT department is its ability to track and monitor all issues, ranging from hardware troubleshooting and software installations to computer/server crashes and network disruptions. The ticketing system effectively organizes and prioritizes the influx of problems, allowing IT staff to navigate the complexities with greater ease and ensuring that critical issues are addressed in a timely manner.

Further, a work order system dramatically improves communication, making it easier not only for IT staff members to stay informed and up to date but also for end-users who might need assistance or clarification. This robust communication platform ensures that no details are lost in translation and that IT managers can keep a finger on the pulse of their department's activities.

The ticket system also acts as a hub for coordinating resources and teams, streamlining the assignment of tasks and allowing for better delegation. By ensuring that IT staff with the relevant skills and expertise are assigned to the right projects, the work order system maximizes team efficiency, reduces the need for micromanagement, and ultimately saves time and effort.

Moreover, work order systems play a crucial role in monitoring accountability and performance. With better visibility into the progress of tasks, managers can more easily identify any instances where a team member may be falling short or exceeding expectations. This insight proves invaluable when conducting performance reviews or making decisions about promotions and rewards.

Additionally, the centralization of data afforded by a work order system produces a wealth of information that can be analyzed and used for improving processes and identifying successful strategies. By examining this data, IT departments can make evidence-based decisions and strategies, leading to better allocation of resources and more targeted improvements in efficiency and productivity.

In conclusion, the importance of a work order system in the IT department cannot be overstated. It supports businesses in managing an array of tasks and challenges, from tracking issues and boosting communication to coordinating resources and analyzing data for continuous growth. By adopting and utilizing such a system, IT departments are well-poised to maintain a streamlined, dynamic, and effective operation that benefits the entire organization.

Project Management

If fortunate, your ticket management will feature a built-in project module. What separates this system from solely relying on stand alone tickets is its capability to record tasks and orders in different stages with its timeline of completion. By having multiple related tickets associated with each stage, the whole process becomes much more manageable and efficient!

A project management system is a powerful tool for any organization, but it can be especially important for an IT department. The benefits of utilizing a project management system in an IT department are many. These benefits include increased efficiency, better communication, and improved collaboration.

There are several benefits to using a project management system in an IT department:

- Increased efficiency using a project management system, where tasks can be tracked and assigned more easily. This helps prevent tasks from falling through the cracks and ensure that projects are completed on time.

- Better Communication is key in any organization, but it is especially important in an IT department. A project management system helps ensure that all team members are on the same page and that information is shared in a timely and efficient manner.

- Improved collaboration using a project management system, allows team members to work together easily and effectively. This helps to improve the overall quality of the work produced and ensure that projects are completed to a high standard.

Examples to highlight the benefits of a project management system for an IT department:

- When a system experiences bugs, it can be difficult to track down the root cause of the issue. Using a project management system, developers can track bugs more easily and work together to find a solution quickly.

- Software development software can be a complex process that involves multiple team members. Using a project management system, allows developers to track tasks and collaborate more effectively.

- Upgrading infrastructure can be a daunting task, but it is necessary to keep systems running smoothly. Using a project management system, IT teams can plan and execute infrastructure upgrades more efficiently.

Tips for Using a Project Management System with an IT Team

1. Set clear goals: Before starting a project, it is important to set clear goals and objectives. This will ensure that all team members are working towards the same goal.

2. Track progress: By tracking progress, team members can see how the project is progressing and identify areas where improvements can be made.

3. Communicate regularly: Regular communication is key to the success of any project. By communicating regularly, team members can ensure that everyone is on the same page and that any issues are addressed quickly.

Success Stories for IT Departments

Microsoft has been using a project management system for many years. By doing so, the company has been able to improve collaboration and increase efficiency.

Google has also been using a project management system for many years. The company has found that the system has helped to improve communication and collaboration among team members.

GE has been using a project management system for several years. The company has found that the system has helped to improve efficiency and reduce costs.

Ticketing System as Project Management

A ticketing system can also be used as a project management tool. By categorizing tickets into groups or categories, it is possible to track progress and ensure that tasks are completed on time. This is especially useful for IT departments that deal with a high volume of support requests. By categorizing support requests, team members can prioritize tasks and ensure that the most important issues are addressed first.

The Inventory System (Module)

As businesses grow and expand, it becomes increasingly important to maintain an organized inventory system to ensure business continuity and success. The IT department plays a vital role in developing and maintaining an inventory system that helps in easy tracking, proper usage, and efficient storage of inventory items.

An inventory system, with proper usage, keeps accurate records of inventory items, which in turn helps with proper budget prediction, depreciation, and accountability. For example, consider a retail store that sells clothing items. An inventory system tracks the clothing items sold, removed, or exchanged, and provides a clear picture of available stock status. Having insight into stock availability can help the retailer to make future predictions, optimize purchasing, and prevent overstocking or out-of-stocking of products.

When selecting an inventory management system, it is important to consider the following items:

1. Cost-effectiveness

2. Scalability

3. Ease of use

4. Access control

5. Reporting abilities

6. Integration possibilities

Remote Management System (RMM) rather than just a remote control system is an essential consideration for any IT department supporting more than fifty workstations and six servers. The features of RMM are notably more expansive, so investing in one is crucial to ensure that your team can provide the best possible support.

Remote management systems offer several key benefits that can help businesses to operate more efficiently and cost-effectively. One of the most important advantages is the ability to manage and maintain devices remotely, without having to be physically present. This allows businesses to quickly identify issues, resolve them promptly and apply any necessary patches or updates promptly.

Remote management systems also allow businesses to automate many of their maintenance tasks, such as running scripts and checks on all devices at a scheduled interval or automatically when certain conditions are met. This ensures that all devices remain up-to-date with the latest security patches and configurations for optimal performance.

Additionally, remote management systems make it easier for IT teams to deploy software applications across multiple devices simultaneously. This reduces manual effort, lowers costs associated with deploying apps on each device, and ensures that all endpoints are configured correctly so they can provide the best user experience possible.

Overall, remote management systems provide significant advantages over traditional methods of device maintenance. From reducing costs associated with deploying applications across multiple devices simultaneously to allowing for faster resolution of issues by enabling users to access endpoints remotely. These systems offer a great way for businesses to optimize their operations while ensuring that all devices remain secure and updated with the latest configurations.

It is important to note that even if you do not have the availability for a sophisticated inventory management system, you can utilize simple tools from office applications such as a simple database at first. This may be a great starting point for small businesses, allowing you to create and manage an organized system for tracking inventory items.

One of the significant benefits of an inventory system is the abundance of data that can be recorded and referenced. Having this data and being able to report on those records by department, employee, or facility is where the strength of such records exists. This feature can aid the business in the decision-making process by providing valuable insights on how to optimize inventory to increase profitability.

In conclusion, having an inventory management system is essential for a business to operate efficiently and effectively. It is the IT department's responsibility to develop and maintain such systems to keep accurate record-keeping of inventory items. Effective inventory management provides a clear picture of stock status, leading to better business decisions in the future. By investing in a comprehensive inventory management system, businesses can remain competitive and maintain success in the long run.

The Document Management System

When it comes to document systems and platforms, there are three distinct types. Each offer something unique and of course pros and cons. Some manage documents through indexing, while others provide simple repositories for storing files or use a combination of both techniques. For those looking to make the most out of their system, options exist that also incorporate repository management and imaging solutions as well- allowing you to do more with fewer resources.

The document management system can effortlessly manage all your organizational complexities and needs, but unless you are operating a massive IT company with a large number of personnel, these features may be unnecessary for achieving success.

Notable Features

1. Secure storage: The best document management systems offer secure storage of documents, with powerful encryption algorithms and access control lists to keep your data safe and secure.

2. Easy search & retrieval: Finding the right document quickly is essential for efficient workflows, so look for a system that offers easy search & retrieval tools like full-text search capabilities.

3. Scalability: The best document management systems provide an optimal level of scalability, allowing you to increase usage as needed without impacting performance or security levels.

4. Collaboration features: Collaboration features like role-based user permissions, version control, and annotation tools are invaluable for effective team collaboration on documents.

5. Integration with other business software: No modern business runs entirely on one platform, so it is important to find a document management system that integrates easily with your other business software like ERP and CRM solutions.

Customizations Document Management Systems

IT teams often do without archiving, retention policies, partial document security, automated movement of documents with workflow control, user management customizations and notifications for follow up purposes or extensive versioning requirements.

Success for Managing Documents

If your IT team is similar to the majority of departments around the globe and consists of fewer than 100 individuals, steer clear of overly complex systems that promise extreme elegance but come along with immense overhead. Steer away from programs with extensive database architecture, intricate complexity for user management and daily operations. These kinds of complicated systems are usually used in large Human Resources departments where tracking workflow processes is paramount.

The IT department requires an efficient password management system and a reliable knowledge base to store policies, protocols, guidelines, and instructions. Select your preferred password manager solution before deploying a WordPress-based or another hosted Knowledge Base that will meet all your requirements.

Visuals are important when it comes to document management. Short videos can be much more effective than lengthy documents, providing employees and customers with a better grasp of the information at hand. To maximize your knowledge base for corporate purposes such as Standard Operating Procedures (SOPs), policies, and standards, try integrating these documents into your "How-To" system - creating an efficient relationship between all three components.

The Setup

By investing the extra effort to build in-house, you can combine all three of these functions into one system: Knowledge Bases (Kbs), "How To" documents, and more. Doing so will save time while also providing a streamlined user experience! Nevertheless, this approach is usually not the first choice for most directors, unless they also manage a development team.

Staying mindful of any regulatory scenarios that your business may be facing is essential. For example, CMMC or NIST requirements could pressure you to use complex systems. However, as long as you abide by their regulations for encryption and secure government-mandated hosting, the format outlined here can still be followed.

Reporting Systems

Reporting tools are invaluable for any IT department as they provide an effective means of tracking progress, managing data, and making informed decisions. These tools allow IT teams to quickly and easily access critical data about projects and performance, enabling the identification of problem areas, tracking of expenses, and measuring team efficiency. Reporting tools can save a tremendous amount of time by eliminating the need for manually compiling reports. Furthermore, these tools facilitate collaboration between teams since everyone can access accurate information in real-time. With the help of reporting tools, IT departments stay organized and on top of tasks while providing managers with an overall picture of how the team is performing. Additionally, the tools enable departments to take timely action if there is an issue or something is not performing as expected. All in all, reporting tools are extremely valuable for any IT department due to the powerful insights they provide, the ease with which they can be used, and the improved decision-making capabilities they allow for.

Many different types of useful reports can be generated from reporting tools for an IT department. For example, a demand analysis report provides insight into the number of users, how often they use resources, and their usage patterns. Another is a project status report which tracks progress on a given project, such as tasks that have been completed or those still in progress. Other reports include budget performance reports to monitor and compare expenses against allocated funds; system inventory reports to track hardware and software assets; and service-level agreement (SLA) reports to measure performance against contractual standards. All these examples offer detailed information about how an IT department is performing, allowing informed decisions about how best to use its resources and allocate its budgets.

LMS (Learning Management System)

A Learning Management System (LMS) is essential for the success of an IT team because it provides a platform for training and development tailored to meet the specific needs of the organization. With an LMS, IT teams can create custom courses and on-boarding materials that are designed to improve skills, increase productivity, and ensure compliance with industry standards. An LMS allows teams to access ongoing support from experts in the field while providing performance metrics to track progress. Additionally, it provides a way for IT groups to collaborate on projects within a secure environment, making communication easier and reducing operational costs. All in all, having an effective LMS is crucial for any successful IT team as it maximizes their potential while minimizing time and resources spent.

An LMS offers a range of benefits for an IT team. Not only does an LMS provide a centralized platform to create, track, and manage courses as well as provide on-boarding materials for new employees but as mentioned earlier, an LMS equips IT teams with needed tools for development and industry standard compliance. Using advanced analytics and performance metrics, an LMS provides detailed insights into employee development, enabling teams to measure progress and identify areas for improvement. This allows for accurate budget allocations and improved resource utilization. Furthermore, an LMS creates ease for IT teams to collaborate through the sharing of ideas, resources, and information on projects within a secure environment. Finally, having access to ongoing support from experts in the field ensures the latest applicable knowledge and best practices which ultimately leads to better results for everyone. In short, an effective LMS is essential for any successful IT department since it provides invaluable tools to help improve productivity while reducing operational costs.

Remote Management Systems

Investing in a Remote Management System (RMM) rather than just a remote control system is an essential consideration for any IT department supporting more than fifty workstations and six servers. The features of RMM are notably more expansive, so investing in one is crucial to ensure that your team can provide the best possible support.

Remote management systems offer several key benefits that can help businesses to operate more efficiently and cost-effectively. One of the most important advantages is the ability to manage and maintain devices remotely, without having to be physically present. This allows businesses to quickly identify issues, resolve them promptly and apply any necessary patches or updates promptly.

Remote management systems also allow businesses to automate many of their maintenance tasks, such as running scripts and checks on all devices at a scheduled interval or automatically when certain conditions are met. This helps ensure that all devices remain up-to-date with the latest security patches and configurations for optimal performance.

Additionally, remote management systems make it much easier for IT teams to deploy software applications across multiple devices simultaneously.

This reduces manual effort, lowers costs associated with deploying apps individually on each device, and helps ensure that all endpoints are configured correctly so they can provide the best user experience possible.

Overall, remote management systems provide significant advantages over traditional methods of device maintenance. From reducing costs associated with deploying applications across multiple devices simultaneously to allowing for faster resolution of issues by enabling users to access endpoints remotely, these systems offer a great way for businesses to optimize their operations while ensuring that all devices remain secure and updated with the latest configurations.

Customer Relations Management System

Customer Relations Management (CRM) is a technology that helps businesses manage their relationships with customers. It is software that stores customer information and interactions in a central database, which can be accessed by employees across different departments. The main purpose of a CRM system is to improve customer service, increase sales, and ultimately grow the business.

The IT department can benefit from a CRM system for internal purposes by using it to store and track internal customer interactions. For example, the IT department could use a CRM system to track employee requests for assistance, such as troubleshooting software issues or setting up new devices. By using a CRM system, the IT department can better manage its workload and ensure that no requests slip through the cracks.

If the IT department services a specific industry like medicine, utilizing a CRM system can provide many benefits for external customers. The system can help the IT department keep track of patient data, schedules, appointments, and invoices, while also providing better customer service. The CRM system allows the IT department to retrieve this information quickly and efficiently, improving patient care and enabling a better experience for healthcare providers and patients.

The pros of using a CRM system include:

1. Enhancing communication with customers
2. Streamlining sales processes
3. Providing a centralized database for customer information
4. Improving customer retention
5. Personalizing marketing communications
6. Increasing efficiency across departments
7. Automating tasks and lowering operating costs

Many work orders/ticketing systems have a CRM as part of their package. This allows businesses to gain the benefits of both systems in one package. A work order/ticketing system with a CRM allows a business to manage its customer interactions, while also tracking work orders and employee tasks in one central location. This improves communication between departments and provide a better overall customer experience.

When choosing a CRM package, there are several points to consider:

Ease of Use
Look for a system that is easy to use and intuitive for your employees.

Integration
Ensure the system can integrate with other software your business uses.

Customization
Choose a system that can be customized to fit the specific needs of your business.

Security
Make sure the system has the necessary security measures, such as two-factor authentication and encryption.

Scalability
Choose a system that can grow with your business and handle increasing amounts of data.

Support and training
Ensure the system comes with adequate support, training, and documentation.

Price
Choose a system that fits your budget, taking into consideration any additional costs such as implementation, training, and maintenance.

A CRM system is an important tool for managing customer relationships, improving sales, and growing businesses. A CRM system can benefit the IT department internally, and be utilized to benefit external customers, depending on the specific industry. The benefits of using a CRM system include streamlining processes, enhancing communication, and providing a centralized database for customer information. Businesses should consider several important factors when choosing a CRM package to ensure they choose one that meets their unique needs.

The Enterprise Resource Planning System

An ERP system (Enterprise Resource Planning) is a type of software application that helps an organization in managing its day-to-day business activities such as accounting, procurement, project management, supply chain operations, and more. ERP systems allow various departments within an organization to share data and collaborate while ensuring that all data is accurate and up to date. This helps streamline business processes, increase efficiency, and make better-informed decisions based on real-time data.

In addition, an ERP system can be used to store and manage business-critical information such as client information, financial data, and other types of sensitive information. By doing so, an IT team can protect data whilst ensuring that only authorized personnel have access. Also, using an ERP system ensures that data integrity and accuracy are maintained consistently as all data is stored on a single platform.

An ERP system can also aid in managing the budget for IT projects. An important aspect of every IT project is budget management. With an ERP system, IT departments can create budgets for projects, monitor expenses, and ensure that they remain within allocated budgets. This not only ensures that projects do not exceed the expected financial limits but also helps in effective financial management.

Moreover, using an ERP system can help IT teams in resource allocation. With an ERP system, IT departments can monitor resource availability such as human resources, software, and hardware, and assign them to the needed projects effectively. An integrated ERP system streamlines the allocation process, making it quicker, easier, and more effective.

The bottom line is that ERP systems can help IT departments in tracking assets and managing critical data, budget management, and resource allocation. Implementing an ERP solution not only streamlines the IT department's operations but also boosts data security and collaboration within the organization. The

IT department can focus on its core function which is to effectively manage and deliver IT services to the rest of the organization, rather than on tedious manual processes.

Words of Caution

CRM, ERP, and PSA are three different software systems that businesses typically use to perform various functions. Although the functions of these systems can overlap at times, they each serve a different primary purpose. CRM software is designed to manage customer relationships, track interactions with customers, and optimize sales and marketing processes. ERP software is designed to help manage business processes such as accounting, procurement, and inventory management. Professional Services Automation (PSA) software is used to manage and optimize service delivery processes.

Although these systems serve different primary functions, they can overlap in some areas. For example, both CRM and ERP systems may have customer-related data, such as contact information, stored within them. Additionally, both CRM and PSA systems may have information related to project management, such as task completion and progress.

When selecting one or more of these systems for your business, it is important to understand the differences between them and to watch for potential overlaps. For example, if your business primarily focuses on providing services to customers, a PSA system may be the most important tool to have. However, it may also be necessary to have a CRM system to manage and nurture relationships with clients. If your business has a large inventory of products, an ERP system might be essential to manage procurement and inventory processes. In addition, as the functionality of these systems overlaps, it is important to avoid redundancy and data segregation.

It is crucial to identify a system that provides functionality in all of the main areas of focus. For example, if your IT team needs a system to manage customer relationships effectively, choosing a CRM system with integrated project management capabilities may be beneficial. If your business primarily focuses on managing products, an ERP system that is designed to integrate inventory management and financial tracking could be invaluable. If you focus on Professional Services as well as product sales, choosing an integrated PSA system could be the right choice, as it provides functionality in both areas.

An example of a company that has successfully integrated these systems is a consulting firm that primarily focuses on project delivery. They use a single PSA system that integrates with CRM and ERP systems to ensure that they have all the necessary functionality to deliver high-quality engagements.

To prevent confusion and maintain clear processes, it is important to avoid implementing multiple systems that perform the same function, duplicate data storage, or overlap extensively in functionality. Instead, choose the most appropriate system for your business needs and ensure that it has the necessary functionality to support all aspects of your operations. Conduct thorough research and solicit input from other stakeholders within your business to ensure that the system you select will meet the needs of your entire organization.

Cybersecurity Systems

These systems help to protect against cyber threats, such as malware, viruses, and attacks on company servers.

In today's technology-driven world, cybersecurity is a top concern for all companies, and therefore IT departments need to invest in cybersecurity systems. A cybersecurity system is a series of tools, techniques, and technologies that work together to protect digital assets from cyber threats such as viruses, malware, and cyber-attacks. IT departments must have a robust cybersecurity system as it ensures the protection of valuable and sensitive data that is transmitted and stored on company servers.

Cybersecurity systems include tools such as firewalls, antivirus software, email security systems, network security devices, and encryption software. These tools work together to help IT teams detect, prevent, and mitigate digital attacks. Some of the most common cybersecurity systems include:

Firewalls
Firewalls act as a barrier between the company's internal networks and external networks such as the Internet. It prevents unauthorized access to the network and restricts traffic from external sources.

Antivirus software
Antivirus software helps to identify and remove malware such as viruses, spyware, and Trojans from company devices. It prevents viruses from infecting the company network by scanning incoming files and emails.

Email security systems
Email security systems help to prevent phishing and spam emails from reaching company inboxes. It also scans incoming emails for potential malware and viruses.

Network security devices
Network security devices such as intrusion detection systems (IDS) and intrusion prevention systems (IPS) detect and prevent cyber-attacks such as denial-of-service attacks and phishing attacks.

Encryption software
Encryption software protects sensitive data by encrypting the data during transit and storage. It ensures that even if an attacker gains access to the data, they cannot read or use it.

Cybersecurity systems are essential for any organization as they provide protection against a wide range of cyber threats that can cause significant harm to the company. Cyberattacks can lead to financial loss, data loss or theft, and reputational damage that can be costly to repair. Investing in a robust cybersecurity system not only helps to mitigate the risk of cyberattacks but also provides peace of mind to the company and its customers.

IT departments should also prioritize education and training to raise cybersecurity awareness among employees. This helps to prevent human errors that can expose the company to cyber threats. IT departments should frequently update their cybersecurity systems and software to stay up to date with the latest security threats and trends.

In conclusion, cybersecurity is a top priority for IT departments, and investing in a cybersecurity system should be a key consideration for all organizations. A strong cybersecurity system not only protects from cyber threats but gives reassurance to the organization and its customers.

Backup and Disaster Recovery

One of the most critical functions of IT departments is to ensure the protection of data and maintain business continuity in the event of a disaster. Backup and disaster recovery systems are designed to safeguard the company's data and recover it in the event of data loss, accidental deletion, or a natural disaster. These systems are essential for protecting against data loss and ensuring that business-critical data is backed up regularly.

Backup and disaster recovery systems work by taking regular backups of the company's data and storing it in secure offsite locations. This ensures that even if there is a hardware failure or a natural disaster such as a flood or fire, the company's data remains safe and can be recovered. The system will back up data at regular intervals defined by the recovery point objective (RPO) and will store the copies in multiple locations.

In contrast, disaster recovery systems enable the prompt and effective restoration of data and technology in case of hardware disappointment, natural catastrophe or cyberattack. To assess how successful these solutions are, it is necessary to consider two essential criteria: Recovery time objective (RTO) and recovery point objective (RPO).

Knowing your RPO is a key step in mitigating data loss during disasters. RPO reflects the timeframe between the last successful backup and when disaster struck, so ideally you should strive for an RPO as short as possible. This will keep any lost data to a minimum, allowing IT teams to quickly restore their systems back up to speed with minimal interruption.

The RTO is the amount of time needed to restore the systems after a disaster. RTO defines the time needed to get back to normal operations, and the enterprise's ability to handle a disaster. A shorter RTO allows the IT department to restore operations faster, reducing downtime and potentially minimizing the company's financial loss.

Backup and disaster recovery systems are essential because they ensure that the company's data is protected and available even in the event of a disaster, reducing the risk of data loss and downtime. They provide peace of mind to both the IT department and the business owners that critical data is continually backed up and can be quickly restored in case of data loss.

Some of the most common backup and disaster recovery systems include cloud-based backup solutions, hybrid backup and recovery systems, and automated backup and recovery systems. IT departments can select the system or systems that are the most suitable for their business needs and fit into their budgets.

In conclusion, backup and disaster recovery systems are key to ensuring data protection and business continuity. IT departments need to have a robust backup and disaster recovery plan in place to ensure that they are always prepared for unforeseen events or disasters. A reliable backup and disaster recovery system should be part of any IT department's arsenal to protect against data loss, and downtime and ensure business continuity.

Collaboration Tools

In today's fast-paced workplace, communication and collaboration are key elements that are essential for success. Collaboration tools such as instant messaging, video conferencing, and team collaboration platforms have revolutionized the way IT teams work together. These tools allow teams to communicate and collaborate more effectively, optimize communication, increase productivity, and simplify collaboration. The following elaborate on the importance of using collaboration tools in the IT department.

Optimize Communication

Communication is essential for the success of any IT department. Collaboration tools offer several communication channels such as instant messaging, voice calls, and video conferencing that foster efficient real-time communication among team members. Remote teams benefit highly from these tools allowing them to communicate and collaborate from anywhere, anytime.

Increase productivity
Collaboration tools streamline the communication process and decrease the time spent switching between various communication platforms. This saves time and improves team efficiency. For example, video conferencing allows team members who work in remote locations to meet virtually, share documents, and work on projects collaboratively in real time.

Simplify Collaboration
Collaboration tools make it easy for team members in the IT department to work together on projects by sharing documents and tools yet maintaining version control. Collaboration platforms allow team members to work together on a project in real time, reducing the time spent on back-and-forth communication.

Foster Creativity and Innovation
Collaboration platforms allow for real-time brainstorming sessions, where team members can contribute their ideas and brainstorm solutions to problems collaboratively. This helps promote innovation and creative thinking within the IT department.

Reduce Email Overload
Collaboration tools such as instant messaging and team collaboration platforms reduce email overload significantly, which is crucial for the IT department. Emails are often a time-consuming method of communication, and team members may be overwhelmed by the volume of emails received. Communication tools reduce the volume of emails while ensuring communication stays efficient.

In the end communication and collaboration are critical elements for the success of any IT department. Collaboration tools such as instant messaging, video conferencing, and team collaboration platforms optimize real-time communication, increase productivity, foster creativity, and innovation, simplify collaboration, and reduce email overload. To reap the benefits of collaboration tools, IT departments must invest in these tools, train their teams to use them, and make them part of their daily communication channels. Collaborative and efficient communication fosters a better work culture and increases opportunities for successful IT projects.

Essential Systems to the Success of an IT Department

After discussing, analyzing, and researching IT tools for a prolonged period, which tools are essential to the success of an IT department?

Maintaining a successful IT department is a complex process that requires careful planning and competent management. The right tools can make all the difference, helping IT professionals work more efficiently and effectively. Below some essential tools and technologies that every IT department should have in its arsenal:

Network Monitoring Tools

These tools track network performance, identify problems and manage network traffic. Some popular network monitoring tools include SolarWinds, PRTG, and Nagios.

Virtualization Software

Virtualization software allows IT departments to create virtual machines and run multiple operating systems on a single piece of hardware. VMware, VirtualBox, and Hyper-V are some popular virtualization software options.

Cloud Services

Cloud services help IT departments save money and improve scalability. Cloud storage options such as Google Drive, Dropbox, and OneDrive are popular options for storing large amounts of data.

Remote Access Tools

Remote access tools allow IT professionals to access computers and servers remotely, which can be useful when troubleshooting problems. Some popular remote access tools include TeamViewer, LogMeIn, and Remote Desktop.

Helpdesk Software

Helpdesk software can streamline IT support processes and make it easier to manage service requests. Popular options include Zendesk, Freshdesk, and ServiceNow.

Antivirus Software

Protecting systems from malware and other cyber threats is essential in today's world. Antivirus software like Norton, McAfee, and Bitdefender can safeguard IT systems.

Backup and Recovery Tools

IT departments need to have a reliable backup and recovery strategy in case of data loss. Backup tools like Acronis, Carbonite, and Veeam can help ensure data is safe and recoverable.

Communication and Collaboration Tools

IT professionals often need to communicate and collaborate with each other and with other departments. Tools like Slack, Microsoft Teams, and Zoom can facilitate communication and collaboration.

These are some examples of the tools and technologies that help IT departments be successful. Every organization's needs will be different, but having the right tools in place can make all the difference.

Build an IT Management Team

The Primary Staff

Building the ideal IT management team requires careful consideration and planning. The first step is to identify core roles within the IT department, such as a Chief Information Officer (CIO) or Chief Technology Officer (CTO). These are the individuals responsible for leading the IT team and setting strategic goals. It is important to ensure that the CIO and CTO have the appropriate background experience and skill sets relevant to the organization to properly manage the team operations.

Once these individuals are identified, it is important to select a team of competent professionals who have the requisite knowledge and experience. Each member of the team must understand both technical and business-related matters so that they can provide effective communication across the organization. Additionally, members should have adequate problem-solving capabilities, organizational ability, and interpersonal skills to resolve conflicts quickly and effectively. Involving members from different departments within the organization in this process can also lead to increased collaboration among different teams when developing IT solutions.

Allocating proper personnel to respective roles and titles is essential for constructing an efficient and effective IT department. Do not make an effort to interchange one role with another or mingle various responsibilities under a single head unless sufficient resources are being provided. Furthermore, actively nurture and develop the skills of your IT team through regular training so that they can keep up to date with new technologies, practices, and procedures. With the right approach to staffing and management, a strong IT management team can ensure the success of any organization.

Finally, it is important to create a culture of innovation within the IT department. This can be achieved by encouraging experimentation through open dialogue between members and providing incentives for innovative ideas or approaches. Creating an atmosphere where new ideas are welcomed, supported, and encouraged, organizations ensure IT teams develop more efficient solutions that benefit both the company and its customers.

Steps for building a successful IT Management team:

1. Identify the needed core roles within the IT department and select an appropriate lead (e.g., CIO or CTO).

2. Choose professionals with technical and business knowledge, problem-solving skills, organizational ability, and interpersonal expertise.

3. Involve members from various departments in IT strategy development.

4. Create a culture of innovation by encouraging open dialogue between members and providing incentives for innovative ideas.

5. Monitor progress regularly to ensure that objectives are being met and adjust if necessary.

Core Roles

The IT department of any organization is integral to its success and longevity. Core roles within the IT department are crucial for organizational effectiveness and staying ahead of industry trends. A strong team is essential for an organization's ability to remain competitive and take advantage of new technology and opportunities as they arise. There are several key roles within the IT infrastructure that ensure a successful operation.

These positions include:

1. Network Administrator: Responsible for managing the network infrastructure and ensuring it is secure from threats.

2. System Administrator: Responsible for maintaining systems such as servers, databases, and applications.

3. Software Developer: Responsible for developing applications and software solutions.

4. IT Security Professional: Responsible for protecting the organization's information assets from unauthorized access or misuse.

5. Data Analyst: Responsible for analyzing data to gain insights into user behavior and business operations.

6. Database Administrator: Responsible for maintaining databases so that they can store, organize, and retrieve data quickly.

Development of IT Strategy

In addition, there must be an appropriate lead position in place to ensure these roles are effectively managed. This could either be a Chief Information Officer (CIO) or a Chief Technology Officer (CTO). A CIO typically oversees the entire IT infrastructure whereas a CTO focuses more on research & development. Both are important positions within the organization as they ensure that the IT operations run smoothly while providing strategic direction.

Ultimately, having strong core roles within the IT department is essential for any successful business. Organizations must select the right personnel who have the necessary skill sets to fill each role, with an appropriate lead at the helm. With this structure in place, companies can feel confident that their IT operations will remain efficient and up to date with current industry trends.

In today's digital world, IT strategy development should involve members from various departments. This is because the strategy must take into account the needs and goals of all stakeholders, not just those in IT. Having input from other departments ensures that your IT strategy will meet the needs of everyone involved, providing a comprehensive solution that benefits the entire organization.

Furthermore, involving members from other departments allows and promotes collaboration within the organization. Working together with people with different perspectives and experiences provides valuable insight into how best to approach various challenges. This creates an environment where ideas are freely discussed, and new solutions are explored. Additionally, it allows for better communication between different parts of the business, helping to foster understanding between teams and improving overall morale.

Finally, bringing in members from departments outside of IT also ensures that any potential risks or conflicts are identified early in the process. This reduces the likelihood of delays in implementing the final strategy due to unforeseen issues further down the line, resulting in a smoother deployment for everyone involved.

In conclusion, integrating members from various departments into IT strategy development is essential for creating an effective plan that meets everyone's needs and ensuring successful implementation throughout the company.

Culture of Innovation

Creating a culture of innovation within an organization is key to achieving long-term success. To do this, it's important to encourage open dialogue. By doing this, you create an environment where employees are empowered to think differently and generate new solutions that can benefit the organization.

First and foremost, organizations should provide training and resources to encourage employees to collaborate on innovative ideas. This could include holding brainstorming sessions or setting up team challenges where teams develop their own solutions for a particular problem. Additionally, make sure everyone has access to the necessary tools and technology they need so that they can work efficiently on any projects they're assigned.

It is also beneficial to recognize and reward employees who come up with innovative solutions. This could be done through awards or other public recognition, but it's equally important to provide tangible rewards such as bonuses or extra vacation days. This will help incentivize employees to continue coming up with creative ideas that drive the company forward.

Finally, organizations should ensure that all levels of staff are included in the creative process. Encourage managers to solicit feedback from their teams, while empowering front-line workers by giving them control at their level over how their jobs are executed. This way all members of the organization feel like they have ownership over what happens in the workplace, creating a sense of unity and camaraderie among colleagues which ultimately leads to higher productivity and innovation across all departments.

In conclusion, creating a culture of innovation requires actively fostering an environment for creativity through collaboration among members in different roles within an organization and providing incentives for innovative ideas. Doing so will result in greater efficiency and long-term growth for your business.

Monitor the IT Management Team's Performance

Monitoring the performance of an IT management team is essential to ensure that objectives are being met and that the team's operations remain effective. Regularly evaluating progress allows for timely interventions, if needed, which can ensure successful project completion.

The first step in monitoring an IT management team's performance is to establish clear goals and objectives with measurable outcomes. This will provide a basis to measure and assess progress. Additionally, it is important to set up key performance indicators (KPIs) such as uptime, customer service resolution times, and more, which provide data-driven metrics on how well the team is doing in terms of meeting their objectives.

Establishing measurable goals helps provide a basis to assess progress. Examples of objectives with measurable outcomes include reducing customer service resolution times from 30 minutes to 15 minutes, increasing the uptime of an IT system from 95% to 99%, improving the response time of an on-line

application from 5 seconds to 1 second, and increasing the accuracy of a machine learning model from 75% to 90%.

It is also important to identify potential issues early before they become problems. This can be done through regular check-ins within the team, making use of customer feedback surveys, and reviewing errors or customer complaints. Finally, communication among all members of the organization must remain transparent so that discrepancies in process or workflow can be addressed quickly and effectively. With these measures in place, organizations ensure their IT teams remain productive over time and continue to deliver successful project

The Team's Structure

Basic Structure

The structure of an IT department typically consists of the following components:

• Systems Administration includes managing hardware, software, and systems to ensure optimal performance.

• Systems Analysis focuses on understanding the needs and requirements of the organization to design and implement IT solutions that meet those needs.

• Data Management and Reporting involves collecting, curating, storing, analyzing, and reporting data related to the operations of the organization.

• Automation involves creating automated processes to improve operational efficiency and reduce manual labor.

• Processes and Documentation includes developing documentation for standard operating procedures (SOPs), policies, and regulations as well as training materials.

• Systems Management responsible for managing hardware, software, networks, storage systems, and other IT infrastructure.

• Business Workflows encompasses tasks such as developing user guides for implementing business processes within an organization's IT system, such as ordering or purchasing.

• Policies and Regulations involve ensuring that all areas of IT are following relevant laws and regulations by setting out clear policies governing the use of systems by users.

• Training involves providing employees with the necessary training in using systems and recommending best practices for secure usage.

• Integration with Other Departments/Vendors ensures that different departments or vendors can interact with each other through a common set of protocols or standards while maintaining security.

Systems Administration

System Administration is the process of managing hardware, software, and systems to ensure optimal performance. This includes tasks such as installing new hardware and software, configuring settings, deploying and maintaining upgrades, monitoring system performance, troubleshooting technical issues, and ensuring security and compliance with relevant laws and regulations.

Some practical examples of System Administration tasks include setting up networks and operating systems on computers, installing applications on machines or servers, configuring firewalls for secure access, creating user accounts for employees to use specific software or services, running backups, and performing regular maintenance checks. In addition, a systems administrator may need to manage IT infrastructure such as virtual machines or cloud services to ensure the stability and scalability of the organization's IT systems.

Systems Analysis

Systems Analysis is an important component of the IT department. It involves understanding the needs and requirements of an organization to design and implement IT solutions that meet those needs. This includes tasks such as identifying areas where existing processes can be improved, evaluating data and analytics to determine technical requirements, developing detailed designs for new systems and applications, testing systems before they go into production, analyzing feedback from users to improve system performance, and optimizing system usage.

Working with other departments within the organization, a systems analyst must ensure that the designed solutions meet business goals and objectives while conforming to relevant policies and regulations. They must also stay up to date on industry trends to recommend new technologies or processes that might benefit the organization. Additionally, their role may involve presenting solutions to stakeholders or providing training to employees on how to use new systems. All this requires excellent communication skills as well as a deep understanding of both operational procedures and technical concepts.

Data Management and Reporting

Data Management and Reporting is a key component of any successful organization. It involves collecting, curating, storing, analyzing, and reporting data related to the operations of the organization. On a day-to-day basis, this may include tasks such as importing external data, transforming raw data into meaningful intelligence, integrating data across disparate systems, designing reports that provide valuable insights on a business's performance, and creating visualizations for easy consumption.

Data Management also requires an understanding of different database technologies such as relational databases (SQL), NoSQL databases (MongoDB), or cloud services such as Amazon Redshift to ensure reliable storage and retrieval of data. Additionally, it is important to understand security protocols to keep sensitive information safe from unauthorized access.

Finally, it is up to the data manager to stay up to date on best practices for managing large amounts of data and applying modern techniques for extracting insights from this data. This requires skills such as Machine Learning and Natural Language Processing to uncover trends and reveal opportunities that can help organizations make informed decisions.

Automation

Automation is becoming increasingly important in the modern world, as organizations look for ways to improve operational efficiency and reduce manual labor. This can involve designing automated processes that handle mundane tasks such as data entry or creating scripts to automate repetitive activities. Task automation also includes building robots and other physical technologies that can be used to perform specific tasks efficiently, as well as leveraging cloud computing to enable distributed computing at scale.

At a deeper level, automation requires an understanding of algorithms and coding languages such as Python or Java. It involves creating software solutions that interact with existing systems with minimal human intervention, often employing artificial intelligence for decision-making or sophisticated pattern recognition. Other tasks may involve analyzing vast amounts of data (big data analytics), rapidly processing transactions (transaction processing), or connecting systems across multiple departments (enterprise integration). All this requires knowledge of automation frameworks and tools such as Selenium or Chef, as well as experience developing secure and reliable applications.

Processes and Documentation

Processes and Documentation are critical for any organization to ensure consistency and compliance. This includes developing documentation for standard operating procedures (SOPs), policies, and regulations as well as training materials that provide clear guidance on how to complete tasks effectively. This helps establish uniformity across the organization, reduce errors, and promote best practices.

The process may involve designing templates for different types of documents such as technical manuals or user guides, writing white papers or blog posts to explain concepts or processes, creating custom forms and checklists, storing relevant data in databases or enterprise systems, creating flowcharts to visualize work-flow processes, or setting up automated reminders to ensure deadlines are met. Documentation should be accompanied by appropriate visuals such as diagrams or screen shots to make it easier to understand complex concepts. The right tools should be used when creating documents - Microsoft Word is often suitable for shorter pieces while Adobe InDesign may be more suitable for longer ones. Finally, it is important to test the documents by providing feedback loops and continual updates to ensure accuracy.

Systems Management is the process of managing hardware, software, networks, storage systems, and other IT infrastructure to ensure it meets the needs of an organization. This requires a wide range of skills such as maintaining and optimizing systems, performing preventive maintenance, installing updates, securing data, and monitoring performance.

Systems Management

Systems management also involves setting up network architectures that balance security and scalability while ensuring availability and reliability. This includes creating redundancies so that if one system fails or an outage occurs, backup systems are in place to take over. It also involves establishing backups for data and logs to protect against data loss or corruption. Finally, it means keeping tabs on system performance metrics to detect declines in performance or any outages before they become critical problems.

The right tools should be used when managing systems - such as monitoring software like SolarWinds or Nagios - but just as important are the people involved in doing this job. Systems administrators must have a thorough understanding of both hardware and software as well as good communication skills so they can effectively collaborate with both users and developers. The ideal candidate should also possess technical knowledge in areas ranging from network architecture to operating systems to cloud computing environments.

Business Workflows

Business Workflows encompass tasks such as developing user guides for implementing business processes within an organization's IT system, such as ordering or purchasing. This requires an understanding of the business process that needs to be automated and the goal to be achieved by implementing it. Once this is defined, it can be translated into a work-flow diagram to better understand and communicate how different processes are connected.

The next step is to develop an efficient solution based on this diagram, considering all the necessary steps required for the process. The solution should automate data collection, provide user interfaces and notifications as needed, and allow for reporting and analysis of results. Finally, another important component of workflows is establishing security protocols, so data remains secure throughout the process.

The right tools should be used when creating work-flows - such as Enterprise Resource Planning (ERP) systems or automation software - but just as important are the people involved in doing this job. Those handling business workflows need to have a deep understanding of the company's operations to create effective solutions, as well as technical knowledge related to relevant tools and programming languages.

Policies and Regulations

Policies and Regulations are an important part of IT management and involve making sure that all areas of IT comply with relevant laws and regulations. This requires setting out clear policies and procedures governing the use of systems by users. This includes access control measures such as authentication and authorization procedures, as well as encrypting data to ensure privacy. It also means developing protocols for data backup and archiving so records can be recovered quickly in the event of a disaster or system failure.

Policies should consider different scenarios that could lead to breaches or misuse of data, such as malicious actors attempting to gain access or employees intentionally or unintentionally misusing information. To make sure everyone is on the same page, these policies should be communicated through user manuals, training sessions, and reminders posted in employee workspaces.

The right tools should be used when creating policies - such as identity and access management solutions - but just as important are people involved in enforcement. Organizations need staff who are knowledgeable about relevant laws and regulations and able to develop appropriate solutions for their specific circumstances.

Training

Training is an important part of any IT system. It involves providing employees with the knowledge and skills necessary to use the system properly, as well as recommending best practices for secure usage. This includes teaching users how to adopt strong passwords, setting up access control measures such as two-factor authentication, and making sure that sensitive information remains protected from unauthorized access.

It is also essential to familiarize users with any applicable laws or regulations related to the system they are using. This helps ensure compliance with industry standards and avoids potential problems resulting from the misuse of data or unauthorized access. Training should also include information about backup protocols in case of a disaster or system failure, as well as cybersecurity measures such as regularly updating software and avoiding suspicious links or websites.

The right tools should be used when training users - such as e-learning platforms - but just as important are the people involved in doing this job. Organizations need staff who understand the technology being used and can explain it in clear language, without oversimplifying it so much that users become careless about security risks.

Integrations

Integration with other departments or vendors is an important part of any IT system. It involves establishing a common set of protocols or standards for different teams to share data, collaborate and interact with each other, while also maintaining security.

This allows organizations to benefit from having a single set of standards across all systems, ensuring both data security and user privacy.

Integration requires strong authentication and authorization measures that ensure that only authorized users can access specific resources. This includes using identity and access management solutions such as Single Sign-On (SSO), two-factor authentication, multi-factor authentication, biometrics, and more. In addition, data encryption can help protect confidential information from malicious actors who might try to intercept it in transit or when stored on a disk.

Organizations need staff who have the necessary technical skills to understand how different systems work together and develop appropriate solutions for their specific needs. As well as technical expertise, having good communication skills is essential for successful integration between teams and external vendors alike. The right tools should be used - such as API gateways - but people are key in successfully integrating different departments or vendors into one secure network infrastructure

Hire the Right Resources

Need to Know

Hiring the right people for IT departments can be a challenging task. It requires more than just technical expertise-you need employees who have problem-solving skills, communication skills, and dedication to make sure they can meet your organization's needs. Additionally, it is important to look for an IT team that has the flexibility to adjust to changing priorities and budgets as necessary. Finding an IT team with all of these qualities can be difficult and time-consuming as you search for the best fit for your company.

What to Look For

1. Technical Expertise: Look for an IT team that has the necessary technical knowledge and expertise to handle your needs.

2. Problem-Solving Skills: A great IT team should be able to identify and solve problems quickly and efficiently.

3. Proactive Approach: An IT team should take a proactive approach to identify potential issues before they become costly problems.

4. Communication Skills: The ability to communicate clearly with both internal and external stakeholders is essential in any IT team.

5. Flexibility: The right IT team must be flexible enough

to accommodate the changing priorities and demands of the organization.

6. Cost Consciousness: It's important that your IT team understands cost constraints and works within budgetary guidelines when making decisions or recommending solutions.

7. Dedication: Ultimately, an IT team should be dedicated to achieving success for the organization through their work.

8. Proclivity for Learning: When seeking out an IT professional, being a fast learner is arguably the most useful asset they can bring to the table -- right alongside their ethical character.

Technical Expertise

When looking for an IT team to manage your organization's technical needs, it is important to look for someone with the necessary technical knowledge and expertise. This includes having experience in developing software, managing databases, deploying networks, or any number of other IT-related tasks. Furthermore, the team should have the ability to analyze complex problems and find solutions quickly and efficiently.

It is also crucial that the team members are knowledgeable about the latest technologies and trends in the IT industry. Keeping up with emerging technologies is essential for staying competitive in this ever-changing landscape. The right technical experts should be able to understand how new tools could be used to improve the efficiency and productivity of your organization.

In addition, the team should be able to create plans on how best to deploy their resources and develop strategies for long-term success. They should not only have experience deploying systems but also have the expertise necessary to ensure these systems are secure and running smoothly while keeping up with changes in the market.

Overall, when selecting an IT team, it is important to make sure they have all of the necessary technical knowledge and expertise needed to ensure your organization's needs are met and all goals can be achieved efficiently.

Problem-Solving Skills

Problem-solving skills are essential for any successful IT team. The ability to identify and solve complex problems quickly and efficiently is key to staying ahead of the competition in the digital world. An IT team should be able to analyze a problem, take into consideration various factors, such as customer needs and market trends, and develop solutions that are both practical and effective.

In addition, an IT team should also possess strong communication skills so they can work with other departments and stakeholders to understand their objectives and find a solution that meets all parties' needs. An experienced IT team should have the capability to articulate their ideas effectively both verbally and in writing so that everyone is on the same page when solving problems.

Finally, an IT team should always be looking for ways to improve existing processes or develop new ones. Having the right people with creative minds who can think outside of the box will help any organization reach its goals faster and more efficiently. Effective problem-solving skills require understanding not only how technology works but also how it affects different aspects of an organization.

Flexibility

To be effective, an IT team must be flexible enough to respond to the ever-changing needs of the organization. By being able to adjust their strategies and operations based on new developments and demands, they can remain agile while ensuring that everyone's needs are met.

The right IT team should also have a proactive mentality when it comes to dealing with potential issues before they become serious problems. This means that they should continuously monitor systems and processes to properly implement updates, anticipate any possible downtime or outages, and prevent malicious attacks. Having contingency plans in place is also essential so that if something does go wrong, they know exactly how to address it quickly and effectively.

Furthermore, an IT team should also collaborate with other departments within the organization to understand different aspects of the business better, as well as identify any areas for improvement or system vulnerabilities that could impact operations. Doing this allows them to stay ahead of any potential problems so that they can be addressed promptly and minimize disruptions.

Overall, having an IT team flexible enough to accommodate changing priorities and demands is essential for successful organizations today. By having a responsive approach toward implementing updates or dealing with system issues along with a proactive attitude when it comes to identifying potential problems, any organization can ensure that its operations run smoothly at all times without any major disruptions or difficulties.

Cost Consciousness

It is crucial that any IT team understands the cost constraints of their organization and works within the budgetary guidelines when recommending solutions or making decisions. This ensures that they can find suitable solutions while also staying within budget, as going over budget can have serious financial consequences for a business.

When it comes to finding solutions, an IT team should take into consideration such factors as ROI (Return On Investment) and TCF (Total Cost of Failure). ROI focuses on the expected gain from investing in a certain solution, while TCF looks at the total cost of any potential failure due to implementing the said solution. By understanding how each solution will impact their budget, an IT team can make better-informed decisions that benefit the organization in both short-term and long-term scenarios.

In addition, it is important for IT teams to be familiar with various technologies available to better understand the different costs associated with them. Not only would this help them weigh the pros and cons before choosing a particular product, but it could also potentially help them identify cheaper alternatives which could save the company money in the long run.

At the end of the day, it is essential that every IT team understands its organization's cost constraints and works within its budgetary guidelines when selecting solutions or making decisions. By doing so they can ensure they are taking cost-effective measures while still providing quality results that benefit both the immediate and future operations of their business.

Dedication

An IT team should be focused on achieving success for

their organization through their work. This requires a comprehensive understanding of the company's mission and operations so that the team can provide solutions that are tailored to its specific needs.

For example, if the overall goal is to reduce costs, then the IT team must consider various actions which could lead to cost savings. Whether it be through improving efficiency or implementing automated processes, they must identify ways to achieve this goal without sacrificing quality or service.

Ultimately, an IT team should be dedicated to achieving success for the organization through their work by taking into consideration its unique goals and needs as well as demonstrating a commitment to staying current on new advancements in technology. Doing so allows them to remain agile while ensuring that everyone's objectives are met at all times.

A Team's Most Valuable Quality

The ability to learn and adapt quickly may be one of the most valuable assets an IT team or an individual can have. In today's digital landscape, technologies are constantly evolving and changing, and if a team cannot keep up with these changes, then they will likely become outdated. An IT team must be able to learn quickly, adapt to new tools and technologies, while staying ahead of competitors. The ability to anticipate future trends and prepare for them is often key to any successful organization.

Another benefit of having an IT team that can learn fast is the agility the organization has when responding to customer needs or market changes. With the right skills, an IT team can rapidly develop new products or services to meet customer demands.

In addition, a team that understands how to work together efficiently is important to ensuring overall success. An experienced IT team should be able to coordinate tasks, delegate responsibilities effectively, and ensure everyone is on the same page when working towards a common goal. Having appropriately skilled personnel with diverse backgrounds who understand the importance of collaboration will help any organization achieve objectives.

All in all, the ability of an IT team to learn quickly and adapt is perhaps one of the most valuable assets in any organization. A quality IT team allows the organization to stay ahead of the competition in today's ever-changing markets.

The Team Does Not Have the Right People

If an IT team does not have the appropriate people in place, the below steps may be taken to remedy the situation:

1. Assess current skills and determine what roles need to be filled.

2. Identify job openings and find qualified candidates.

3. If needed, hire new employees or outsource specific tasks to contractors .

4. Train existing staff in any necessary skills that are lacking.

5. Establish clear job descriptions and expectations for all team members.

6. Monitor performance and ensure each team member is meeting goals and objectives regularly.

7. Create a collegial environment that promotes collaboration between departments and encourages creativity within the organization.

The ideal IT professional is highly knowledgeable and highly skilled in a variety of areas. IT personnel should have a strong understanding of computer systems, networks, software development, databases, cybersecurity, and other related technologies. Additionally, personnel should be adept at problem-solving, finding creative solutions to difficult situations which may arise.

An ideal IT professional should also possess strong communication skills in order to effectively interact with customers and colleagues alike. This includes being able to explain complex concepts clearly and succinctly in both verbal and written form. Additionally, they must be able to lead projects confidently while motivating others.

Furthermore, the ideal IT professional must stay up-to-date on emerging trends within the industry. This permits them to provide relevant advice as well as suggest potential plans of action that could benefit their organization.

In summary, the ideal IT professional combines technical knowledge with interpersonal and communication skills to ensure success for not only themselves but for their organization as well. Striving for excellence by staying ahead of the curve and constantly learning new information to best serve those around them completes the ideal IT professional.

Best Practices for Hiring IT Talent

Finding the right people to fill IT roles is critical to success in today's tech-driven world. Best practices to ensure you hire the right tech talent for your organization are:

1. Know Your Needs: It is important to have a clear understanding of what your company needs from its prospective IT employees. Defining these needs should be one of the first steps taken when beginning the

hiring process.

2. Thorough Screening Process: Once you have identified your specific requirements for the position, make sure you have a thorough screening process in place for candidates. This should include conducting background checks, assessing technical certifications and qualifications, interviewing, and conducting skills tests where applicable.

3. Showcase Company Culture: All candidates must gain an understanding of how they will fit into the organization's culture if hired. Make sure to be open and honest with potential hires about what it's really like working at your company and provide plenty of opportunities for them to meet with existing staff members or learn more about the company during their interview process.

4. Compensation Matters: Competitive compensation packages are essential when competing against other organizations for top-tech talent. While money is not everything in terms of attracting good talent, it is important to consider when creating an attractive offer that balances salary expectations along with other aspects such as flexibility, benefits, among other areas of compensation.

5. Never Stop Learning: Keeping up to date on current trends within the industry helps organizations remain competitive and attract top talent who are up-to-date on their skillset and technology knowledge too! Learning new technologies and offering training opportunities can go a long way toward finding qualified job seekers who can confidently tackle any project thrown their way.

These best practices help any organization effectively identify and hire great IT professionals who can help propel them toward success in today's highly competitive environment!

A Guide to Developing an Effective Interview Process

The interview process is a critical step for any organization when hiring new employees. Some tips and guidelines to help ensure your organization develops an effective interviewing strategy:

1. Define Your Objectives: Before beginning the interview process, it is important to define exactly what you are looking for in terms of skills, experience, and abilities. Creating job descriptions that clearly outline the needs of the position can help guide your decision-making throughout the hiring process.

2. Determine Your Questions: Once you have a clear understanding of what type of candidate you need, come up with questions that will provide insight into their background, skill set, and experience related to the role they would be filling. It is also important to consider diversity as part of your question selection—asking about interests and hobbies is a great way to gain insight into how candidates may fit within your company's culture and values.

3. Set Up an Interview Structure: It is important to establish a structure prior to each interview session so that everyone involved knows exactly what the expectations are for each meeting. This should include information on how long the interviews will last, who will participate in each session (including roles), where it will take place (in-person or virtual), etcetera.

4. Take Notes During Interviews: Investing time in taking notes during each interview can help distinguish one candidate from another when it comes to the time for decision-making later. This should include not just basic information such as qualifications but also

personal detail like demeanor or communication style so you can make more informed decisions briefly in comparison with other candidates being considered for the same role.

5. Ask for References: Finally, do not forget to ask each candidate for references. Contacting any previous employers or colleagues can be a great way to gain additional insight into characteristics such as work ethic or team working ability that may not have been visible in their initial interview with your organization—and could help inform final decisions about which candidate makes the most sense for any given role.

Tips for Interviewers:

- Remain professional throughout every session.
- Assess both hard skills and soft skills when determining potential candidates.
- Give all applicants equal opportunity by asking similar questions across interviews.
- Pay attention to both verbal cues and body language when assessing how well someone may fit within your organization's culture.
- Remember that there is no one "right" answer—it is important to assess if someone has potential rather than focusing only on technical qualifications.

Interviewing: What Not to Do

The interview process can be a make-or-break moment for potential job candidates. As such, it's important to ensure that your team knows what not to do when interviewing. Some tips to keep in mind during the interviewing process:

1. Do not Rush the Process: You must take your time throughout the interview process to assess qualifications effectively and properly evaluate each candidate's suitability for the role. Rushing through this

step could lead to poor decision-making or overlooking qualified candidates in favor of those with more impressive resumes on paper.

2. Do not Ask Unnecessary Questions: During your initial conversations with each prospective employee, it is important that all questions directly relate to their qualifications or ability to fulfill the job requirements. Avoid asking questions about personal situations or other unrelated topics as these can put candidates off and make them feel uncomfortable.

3. Do not Take Advantage of Power Dynamics: The interviewer-interviewee power dynamic can sometimes create an environment where an employer takes advantage of a potential hire by not providing information or explaining answers thoroughly enough. It is important to always remain professional and courteous during interviews so that everyone involved feels respected and at ease with the situation.

4. Do not Assume You Know Everything: While it is important for employers to present their expectations and needs up front, it is also critical for them to keep an open mindset throughout the hiring process as well. Remember that there are always different perspectives on any given issue, so do not presume you know exactly what kind of person will make the best fit for your organization until after you have completed your research into possible applicants.

5 Do not Make Decisions Solely Based on Resume Qualifications: While qualifications may be one factor when considering potential employees, remember that oftentimes it pays dividends to take a chance on underrated talents who may have less experience but show promise in other areas such as problem-solving, creative thinking, collaboration, or ambition.

Following these tips when interviewing potential hires,

ensures employers in making better-informed recruiting decisions while also treating all participants fairly and respectfully throughout the selection processes.

Build a Team

Team Building is Vital

Team building is a vital element of any successful business, as it helps to foster collaboration, build trust and strengthen relationships among employees. Not only does team building create a positive working environment, but it also brings many benefits for the company, such as improved communication and problem-solving skills, increased productivity and creativity, better workplace morale and higher job satisfaction.

Team building activities come in many forms, from team meetings and group exercises to fun activities like scavenger hunts or escape rooms. These activities help employees learn to work together more effectively by developing social skills and understanding one another's strengths and weaknesses. They also make them feel connected to each other while boosting morale.

For employees, team building provides an opportunity to get to know each other outside of their day-to-day roles at the office. It promotes collaboration by forming connections between people that may not have had the chance otherwise. Team building also allows employees to practice problem solving in a low-stakes environment which helps them learn how to think quickly on their feet when faced with challenging tasks or difficult conversations at work.

Studies have shown that team building activities can increase employee engagement which leads to improved performance and productivity overall. When team members are united around a common goal and understand one another's strengths/weaknesses, they are better equipped to tackle problems in a more efficient manner. Furthermore, research has found that teams that engage in regular teambuilding activities are more likely to demonstrate cooperation rather than competition when completing tasks together.

Overall, investing in team building is essential for any organization looking for long-term success; it helps create an atmosphere of trust and respect between employees while unlocking the potential of teams through collaboration and creativity!

Team vs IT Team

The IT department is an integral part of any organization, responsible for providing technology solutions, support, and maintenance. As such, many traditional team-building activities may not be suitable for this type of team.

IT teams usually consist of tech-savvy and creative workers who think outside the box to solve complex problems. Thus, it is important to tailor team building activities to their specific needs - one size does not fit all when it comes to IT teams. For example, instead of icebreakers or group games that encourage people to get to know each other better in a fun environment, IT team building activities could focus on problem solving by allowing members to brainstorm solutions or develop prototypes for projects.

Another difference between traditional team building and IT team building is the way goals are set and achieved. While traditional teams tend to follow a more linear approach to goal setting - with clear objectives and milestones - IT teams often take a more iterative approach since they're working with technologies that are constantly changing and evolving. This means that the goals may be less clear cut, but still involve reaching certain milestones over time until the project is completed successfully.

Team building within the IT department can be quite different from traditional team building due to its unique context and technical challenges. However, with effective planning and creative solutions tailored specifically for your IT team's needs, you can create a positive and collaborative atmosphere where everyone can thrive!

Think About How To Do the Following:

- Create an open and inviting atmosphere in the workplace - Encourage collaboration by creating a meaningful dialog between team members and giving everyone chance to share their ideas.
- Promote positive reinforcement - Praise team members for their achievements and encourage healthy competition among team members so that they stay motivated and become more productive.
- Foster trust, understanding and respect - Support each other's efforts, listen to different perspectives, and celebrate successes together as a team.
- Set clear goals - Establish objectives for the team and provide resources to help them achieve those goals quickly and efficiently.
- Take time to recognize contributions - Acknowledge individual contributions, as well as the collective work of the team efforts, to keep everyone united and focused on their tasks.

Accountability

Accountability is an important part of team building, as it encourages collaboration and ensures that everyone is pulling their weight to achieve the desired outcome. When working in teams, it is essential to maintain standards of conduct and uphold certain expectations so that individuals remain responsible for their actions. Guidelines to enforce accountability while building a team:

• Set Clear Goals and Expectations: Make sure that each person on the team understands what they need to do in order to reach the target goal or solution. Provide detailed instructions, timelines and deadlines if necessary.
• Establish an Open Dialogue: Encourage open communication between team members so that everyone feels comfortable expressing their ideas and questions without fear of judgment or criticism.
• Assign Responsibilities: Assign specific tasks to each person on the team so that there's no confusion about who's responsible for what part of the project. This will help ensure that everyone is contributing equally and staying accountable for their work.
• Track Progress: Monitor progress regularly by having regular check-ins with each member of the team – this ensures that any issues can be addressed quickly, before they become major problems.
• Give Constructive Feedback: Provide timely feedback to individuals when needed - both positive feedback when someone succeeds and constructive advice when they could improve - but make sure you keep things friendly and professional! This way, people feel motivated to stay on track and do their best work at all times.

By following these steps, you can foster a sense of responsibility within your team and ensure that everyone remains accountable for their work – which is key for any successful project.

Morals

Morals are an integral part of team building. When people come together as a team, they need to have faith in one another and respect each other's values and beliefs. This trust and understanding create a stronger bond among team members, which leads to greater collaboration, improved communication, and more successful outcomes.

Having morality within a team helps create healthier relationships and foster a productive environment where everyone feels safe to express their opinions without fear of judgment. It also encourages creativity and fosters teamwork, allowing for better problem-solving skills and novel solutions. A strong moral compass will help remind the members that regardless of their differences, they are still working towards the same goal: success.

When leaders demonstrate strong morals in the workplace, it sets an example for the rest of the team to follow. They can form expectations on how others should conduct themselves when working with one another and cultivate an atmosphere of trustworthiness, respectfulness, accountability, honor, reliability, and kindness. That type of culture encourages others to take responsibility for their actions and remain devoted to the group's objectives while maintaining integrity throughout all processes.

It is essential that teams have clear boundaries when it comes to moral issues that might arise during team activities or meetings such as language usage, appropriate behavior or inappropriate contact between members outside work hours. This allows for healthy discussions about controversial topics without creating bad vibes or tensions among employees who may feel uncomfortable with certain views being shared due to personal beliefs or backgrounds.

Ultimately promoting moral conduct within teams has many tangible benefits such as helping individuals bond quickly with one another; improving communication among peers; eliminating gossiping; yield better results from collaborations; generating innovative ideas through mutual respect; and establishing strong leadership roles. Therefore, inculcating these positive traits during team building can be highly beneficial for your organization in terms of productivity levels at workplace along with amazing opportunities for endless growth.

Dealing with Team Failures

It is only natural to feel discouraged after a failure—it is a part of life. When dealing with failure within an IT team, it is important to remain focused on the task and learn from the experience. Tips on how to deal with failure within an IT team:

1. Acknowledge mistakes. Before any solutions can be made, it is important to take responsibility for the mistakes and failures that occurred. Acknowledging them openly will ensure that the team is on the same page and can move forward in learning from their experiences.

2. Assess what went wrong. Once you have taken responsibility for the errors, brainstorm as a team to identify what went wrong and why. Did you not have all the necessary information or skills? Did you overlook something? Was there an oversight or communication breakdown? Understanding these issues will help prevent future problems as well as help solve current issues.

3. Create strategies for improvement. After identifying

where things went wrong, create strategies for improvement, both prevention strategies and repair strategies if needed, to prevent similar issues from happening again in the future. This includes analyzing software, data security practices, processes, and creating protocols to guide future behavior to avoid repeating mistakes while increasing overall success rates.

4. Re-evaluate resources and goals- It is also important to re-evaluate your resources – both human capital and physical asset needs – as well as your overall goals when dealing with failure in team settings yourself what new skills need to be acquired or developed to ensure success in the future? What should our goals be? Identifying these areas of improvement can help set your team up for long-term success instead of short-term gains that may lead back to previous mistakes.

5. Utilize past successes- Remembering past successes and replicating successful processes can also help boost morale after failure Evaluating prior successes will demonstrate potential achievable results even during times of difficulty - this will give employees higher motivation and hope towards reaching success no matter how challenging it may seem.

Overall, dealing with failure within an IT team is never easy but it does not have to be daunting if following these five steps - acknowledging mistakes, assessing what went wrong, creating strategies for improvement, reevaluating resources and goals, utilizing past successes you can work towards improving processes within your team to achieve greater success moving forward.

Things to Avoid

When building an IT team, there are certain pitfalls you'll want to be aware of to create a successful and productive team. Five tips for avoiding common mistakes when constructing an IT team:

1. Do not overlook team dynamics: It is important to be mindful of how members of the team interact with each other, as this can have a major effect on productivity and job satisfaction. Make sure that everyone is working together effectively and efficiently by fostering a positive environment and promoting open communication.

2. Avoid unrealistic goals or expectations: When setting objectives and expectations for your IT team, make sure that they are realistic and achievable. Setting unattainable goals can result in frustration and potentially lead to disengagement amongst members of the team.

3. Ensure adequate training is provided: Providing employees with sufficient resources and guidance will help them gain the skills needed for their roles, reducing the number of errors made throughout projects. It is also important that ongoing training is offered so that employees stay up to date with all relevant technologies.

4. Do not forget about job satisfaction: Investing in employee happiness should also be a priority when constructing an IT team, as satisfied workers tend to produce better results than disgruntled ones it is important to factor in shared interests, working styles, prior experiences, etc., when putting together an IT Team so that everyone feels comfortable working together

5. Designate clear roles and responsibilities: Finally, every member of the IT Team must know the expectations at any given time. In order to do this properly each person needs designated responsibilities, and take ownership. This should include clarity on reporting lines as well so it is easier for people to communicate information quickly

Overall, building a successful IT Team is not easy - but making sure you do not fall into any common traps will give your organization its best chance at success.

Summary

1. Define Goals: Establish clear goals and objectives that everyone can work towards.

2. Set Expectations: Make sure that all team members are on the same page when it comes to expectations, roles, and responsibilities.

3. Build Trust: Create an open and trusting environment where team members are comfortable expressing themselves and sharing ideas.

4. Provide Support: Make sure each team member is supported and encouraged in their efforts, no matter the level of experience.

5. Hold Regularly Scheduled Meetings: Ensure that clear communication lines are established so that everyone is updated on progress and any issues that may arise during a project or initiative.

6. Celebrate Achievements: Remember to take time to celebrate successes as they come - small wins add up over time.

How to Manage Request for Proposals

What is a Request for Proposal?

A Request for Proposal, (RFP), is an invitation to vendors to submit a formal proposal in response to a specific request. It typically follows a lengthy process that includes researching and analyzing qualifications and capabilities of potential vendors, setting detailed criteria, evaluating the proposals received, and selecting the most suitable vendor. There are best practices when obtaining an RFP:

Clearly define your needs

When attempting to define your needs for an RFP, it is important to consider the scope and complexity of the project, how long it will take to complete, the resources required, and any specifications or criteria that must be met. Before releasing an RFP document, you should have identified the exact type of services or products necessary to successfully complete the project. This includes understanding the types of vendors needed and what level of expertise they possess in relevant areas. It should also be determined if there are other factors that need to be taken into consideration such as geographical location or cost considerations. Once these elements have been identified, create a list of criteria for evaluating potential vendors. These criteria should include not only technical qualifications but also non-technical elements such as pricing and the capability of delivery. Finally, by thoroughly researching your options before sending out an RFP document, you can narrow down your selection process and make sure that you are selecting a vendor that meets all your needs.

Set Comprehensive Evaluation Criteria

When setting comprehensive evaluation criteria for an RFP, it is important to consider both technical and non-technical elements. This ensures that the organization requesting the proposal gets a balanced view of all responses. On the technical side, evaluation criteria should include qualifications such as relevant experience in the specific field, level of expertise required for the project, technological capabilities, and any other relevant certifications or accreditations held by the potential vendors. As far as non-technical criteria are concerned, these may involve aspects such as price and timeline for completion of services/delivery. In addition to this, there may be other factors that must be considered such as geographical location or ethical considerations. By making sure that all potential vendors are evaluated according to a well-defined set of criteria, you can ensure an objective selection process.

Communicate Effectively with Vendors

When communicating with potential vendors while processing an RFP, it is important to ensure that all parties have sufficient information and clarity throughout the bidding process. To do this, establish clear guidelines and timelines for each step of the process. This includes stating the desired outcomes or deliverables as well as providing detailed instructions on how to submit a proposal, who can respond, what information is needed from the vendor, and when responses are due. Additionally, all communication should be in writing so that there is a record of everything being discussed and agreed upon. Having defined guidelines and communication expectations in place before releasing an RFP document, ensures that each party involved has the information they need in order to respond effectively.

Closely Monitor Responses

In order to monitor incoming responses to an RFP, it is important to have a tracking system in place. This may involve using software or tools that can manage the data associated with each submission so that any differences between bids can be easily evaluated in comparison with other submissions. Additionally, having a secure internal process for logging and rating the various proposals received ensures that each bid is given the same level of consideration. Furthermore, if issues arise during reviewing responses, having a monitoring system in place will ensure they are quickly and effectively addressed while still allowing sufficient time for all vendors to submit their proposals. In this way, you can ensure that all vendors are treated fairly throughout the entire selection process.

Negotiate Effectively with Selected Vendor(s)

When it comes to negotiating effectively with the selected vendor(s), it is important to have a clear idea of what your key objectives are before entering into negotiations. This means understanding what you want out of the deal, including pricing, terms and conditions, timeline for delivery, payment terms, and any other details that may be relevant to the agreement. After assessing all available bids, open negotiations with the chosen bidder regarding these points to come to an agreement that is mutually beneficial for both parties. Once all points have been discussed and agreed upon, make sure to get everything in writing before signing any contract agreements. In this way, you ensure that all parties involved are satisfied with the result while still adhering to the spirit of fair bargaining.

Things to Avoid when Negotiating an RFP

When negotiating an RFP, there are some common pitfalls to avoid in order to ensure a successful outcome. First, never underestimate the importance of proper preparation. Make sure to thoroughly research all potential vendors and their offerings before entering into any negotiations. Also, approach negotiations with realistic expectations as well as a good understanding of what you're trying to get out of the deal. Additionally, be aware of potential differences in approach and sensitivity levels between different cultures when negotiating with international partners. Finally, do not get emotional or aggressive during negotiations, as this could lead to less than satisfactory outcomes for both parties involved. By following these simple steps, you can ensure that your RFP negotiation process is fair and effective for all involved.

Skills a Negotiator Should Possess

Empathy: Being able to read and understand how the other party is feeling and adjusting accordingly is an important part of successful negotiation.

Patience: RFP negotiations can often take a long time, so one must remain patient throughout the process.

Analytical Thinking: Negotiators should be able to analyze complex data and identify potential solutions or compromises.

Conflict Management: Successful negotiators are skilled at finding common ground and managing any disagreements in order to reach a mutually beneficial outcome.

Flexibility: An effective negotiator should have the ability to pivot quickly between ideas in order to find the best possible solution for both parties involved.

Research Skills: It is crucial that negotiators understand their own interests as well as those of their counterpart in order to accurately assess the situation and craft effective solutions when negotiating an RFP.

Manage IT Like a Business

Why it Matters

For businesses to thrive in today's competitive environment, prominent entrepreneurs advise that you should make ordering as convenient as possible, provide reasonable prices for your products or services, and ensure prompt delivery. IT departments that prioritize fair prices, quick delivery, and high-quality services will be more likely to succeed in the long run and overcome any obstacles they may face. This is the formula for success.

Benefits of Managing your IT Department as its Own Business

Managing your IT department as its own business offers a multitude of advantages. By streamlining operations, you can reduce costs, increase productivity and efficiency, and drive innovation. You can also benefit from having greater control over the services and systems you provide to customers, which ultimately gives them a better experience.

Furthermore, by leveraging the latest technologies in cloud computing, analytics, automation, and digital transformation initiatives such as DevOps, artificial intelligence (AI), and machine learning (ML), your IT department can become an agile organization that responds quickly to customer needs and trends. This ability can give your business a competitive edge in an increasingly technology-driven world.

By establishing an IT department as a business entity within your company, you can create a centralized hub for managing data and information technology resources. You will be able to more accurately measure performance across IT teams; determine which systems are performing best or need improvement; identify areas where new investments should be made; and prioritize areas requiring additional staffing or resources. Additionally, data security will be strengthened by having clear roles and responsibilities laid out for who is responsible for what's going on in each area of the system.

Managing your IT department as its own business opens up opportunities to explore new partnerships with vendors and developers that could lead to increased revenue streams or improved efficiency. By forming strategic alliances with these partners, you could gain access to emerging technologies while introducing professional development opportunities that strengthen confidence in the IT staff's ability to deliver quality solutions promptly.

Managing your IT department as its own business is likely to result in tangible benefits for both customers and stakeholders alike. From cost savings to improved service delivery or efficiency gains through collaboration with outside partners - there are numerous ways this approach will help set you up for success now and into the future.

Keeping it Simple

The typical IT department is accustomed to receiving requests and prioritizing them based on the status of an individual within the organization or a priority set by leadership. Deadlines may not be consistent, and it can often become a race for first-come-first-serve service. Unfortunately, this model does not lend itself to success - something must change!

The typical IT department model of responding to requests by prioritizing them based on the internal hierarchy of an organization or according to leadership is inefficient and ineffective. Such a system can led to chaotic conditions, with deadlines shifting and resources being pulled in different directions. This lack of consistency often leads to unsatisfactory results, with tasks not completed or delivered late.

To produce quality results, the typical IT department should transition away from this traditional method of operation. Instead, they should focus on establishing reliable processes and timelines for each task so that all resources are managed efficiently and effectively. By implementing such a system, IT departments can ensure that projects are completed on time and within budget.

It is also important for IT departments to create service level agreements (SLAs) with their stakeholders to guarantee expectations are met and deliverables are provided promptly. Doing so provides structure and clarity, ensuring that IT teams know exactly how long each task should take and what the deliverables should look like upon completion. It also helps prioritize tasks based on their importance - allowing more urgent tasks to be addressed first while less important ones are put off until later if necessary.

Finally, the IT department needs to take a proactive approach when dealing with requests by engaging early with stakeholders to understand their needs better and provide appropriate solutions quicker. Doing so facilitates collaboration between all parties involved, leading to better-informed decisions being made faster without compromising quality or productivity levels.

Overall, transitioning away from the traditional reactive approach employed by many IT departments is paramount in today's fast-paced business world. By establishing SLAs, creating reliable processes and timelines, and taking a proactive stance when engaging with stakeholders - organizations can ensure that their IT departments are well-equipped for success both now and well into the future.

Reliable Processes for Success

Fast and efficient work orders (Tickets): An efficient system for receiving work orders is essential for any IT department to ensure that each task is recorded accurately and on time. This is true regardless of the source of the request - whether it's verbal, digital, or automated.

Having an organized approach allows IT teams to capture all necessary information to begin work on a task quickly, thereby reducing time-consuming back-and-forth communication with requesters. By gathering the right information upfront, you can set yourself up for success by having a clear scope of the project before any costly missteps are made.

To make sure that each work order is documented properly and in a timely fashion, it's important to have well-defined procedures. These should encompass every step of the process from receipt to assignment (including validation), as well as specific methods for how requests are processed depending on their origin and seriousness or urgency. When it comes to processing digital requests in particular, having an automated workflow is ideal; this ensures that all data points are captured accurately and transmitted quickly and securely.

Moreover, when establishing your system for receiving work orders, it is essential to consider how easily new tasks can be added over time so that no request falls through the cracks. A centralized database or repository can be used to store all inquiries which come in from various sources - allowing teams to easily track progress on multiple tasks at once and view historical data related to customer requests.

At its core, having a system in place for receiving work orders helps simplify the entire process by streamlining operations while ensuring that all activities are tracked accurately and efficiently. It also allows IT departments to respond faster while providing everyone involved with better visibility into what's going on with each project - giving them more control over outcomes and increasing their chances of delivering successful solutions overall.

Control Cost

Understanding the scope and core of a business is key for any IT department when it comes to controlling costs. It is important not to simply spend money on expensive systems that may not necessarily increase performance or efficiency. While you certainly do get what you pay for, an educated decision should be made regarding solutions and processes rather than relying solely on marketing literature.

Researching the most effective path that allows for both efficiency and costs can help ensure that great solutions are delivered at even greater prices. This will create an environment where long-term success is achievable even with limited resources. Considering the current scope of the company and its operations, as well as taking the time to research which solutions would be best, can bolster progress while keeping costs low.

Fast Delivery

Implementing good systems and processes alongside strong company policies can be a key factor in the successful execution of work orders. It is important to ensure that these procedures enable efficient and timely completion of tasks, as this will lead to greater overall productivity for the organization. Establishing clear protocols and expectations can help foster a sense of urgency and accountability among all employees, from managers to those on the ground level. Additionally, investing in automated solutions like software or robotics can greatly reduce the amount of manual labor required while streamlining the process further. This will not only free up resources but also minimize potential errors that may be caused by mundane tasks being performed manually. When equipped with good systems, processes, and strong company policies, organizations have a better chance of executing tasks on time with fewer delays.

Providing a fast service is essential for an IT department; it allows employees or customers to be more satisfied with the services they receive and boosts the overall reputation of the company and the IT department. Quick response times can lead to better customer or employee retention, as people are more likely to stay loyal to companies that offer prompt support when they have questions or issues. Fast service also enables organizations to make better decisions more quickly, as they don't have to wait around for responses. This helps them keep up in competitive markets and remain ahead of the curve. Ultimately, providing a fast service can create a vital competitive advantage and help companies reach their short-term and long-term goals faster than their competitors.

Summary

- Streamline the process for employees and customers to submit work orders, simplifying their daily operations.
- Optimize systems and processes to cost-effectively maximize results.
- Aim to complete orders with the speediest possible turnaround time.

Paralysis From Analysis

Do Not Complicate Unnecessarily

As IT professionals, it is essential to remember that our natural inclination may be toward creating overly complex solutions. To run an efficient department, we must learn to restrain ourselves and instead focus on finding simple yet effective resolutions. By providing clear guidance and direction, we can reign in the collective impulse for convoluted thinking - leading us down a path of reasonability and efficiency. Over complicating the IT department can lead to undesirable results.

Poor network topologies and practices are to blame for many of these issues. For example, not taking into consideration plans when initially setting up a network can result in major headaches down the line when trying to scale or upgrade. Having too many layers of subnets and access control rules in place can create confusion for both admins and users, leading to multiple time-consuming troubleshooting sessions. Not properly configuring firewalls, routers, and other security protocols can lead to major security vulnerabilities, resulting in data breaches and significant financial losses.

Similarly, having an inefficient network architecture or misconfigured equipment can significantly decrease efficiency and performance due to wasted resources. In large installations, it is common for teams to use outdated systems or neglect proper maintenance which leads to degraded services and increased downtime. This often has a direct impact on the business's bottom line as customers become dissatisfied with poor performance or reliability.

The solution is simple: design networks with simplicity in mind while ensuring optimal performance, flexibility, scalability, security, and maintainability. It is also important that the chosen hardware meets the latest standards so that the company doesn't fall behind its competitors in terms of speed and efficiency. Finally, effective backups must always be created so that any sudden disruption does not cause long-term damage or service outages.

Over complicating IT can easily lead to "paralysis from analysis" in organizations, defined as endlessly delaying decisions due to overthinking and over-analysis. This situation is caused by the excessive complexity of processes and systems, resulting in a state where individual tasks take too long to complete which ultimately harms productivity and efficiency. For example, many companies fall into the trap of trying to build a perfect system that meets all their current needs, while failing to plan for future growth or external changes.

For businesses to avoid these pitfalls, they should always keep simplicity in mind when designing their networks while ensuring optimal performance, flexibility, scalability, security, and maintainability. Additionally, the chosen hardware must meet the latest standards so that businesses do not fall behind their competitors in terms of speed and efficiency. Finally, effective backups should be created so that any sudden disruption does not cause long-term damage or service outages.

Outlining Solutions for the IT Department

1. Define the Problem: The first step when outlining solutions for the IT department is to clearly define and identify the problem that needs to be solved. Think about what areas need improvement, or what tasks need to be completed, and formulate a clear definition of the problem.

2. Research and Analyze: Once you've identified the problem, it's important to research best practices within the field and analyze potential solutions. Look at existing technology and software that could help address the issue and consider if any changes or new initiatives should be implemented.

3. Develop a Plan: Using your research and analysis as a basis, develop an action plan detailing how you will address the issue. Break down each step into smaller goals that can easily be tracked, considering external factors such as time constraints or budget restraints.

4. Implement a Solution: With your plan in hand, it is time to put it into action. Utilize existing resources to efficiently implement your proposed solution, adjusting as necessary along the way.

5. Evaluate Progress: As you go through with implementation, take regular stock of progress made towards solving the identified problem. This will help ensure that resources are being used wisely and objectives are being met on schedule.

6. Troubleshoot Issues: No matter how much planning is done beforehand, issues can still occur during the implementation of any solution – this is where troubleshooting comes in. Pay attention to any bumps encountered along the way to improve future processes so they remain effective and efficient going forward.

Tips for Keeping Your IT Department Rolling with Effective Solutions

1. Keep your staff up to date on the latest trends, software, and hardware solutions. Make sure everyone understands how to use them effectively and stays ahead of potential challenges.

2. Leverage the existing skill sets of your employees. Utilizing their knowledge and expertise will help bring new ideas to the table that may provide unique solutions to common problems.

3. Take advantage of available resources such as online forums and tutorials when addressing tricky technical problems. These can be incredibly valuable when trying to find solutions quickly.

4. Automation can do wonders in helping keep your IT department running smoothly. Whether you are automating time-consuming manual processes or utilizing AI-based tools, there are plenty of ways to streamline operations while ensuring accuracy and efficiency simultaneously.

5. Establish effective communication protocols between teams so that all members know what is expected to remain productive and organized during problem-solving sessions or meetings discussing new projects or initiatives.

6. Develop a comprehensive cybersecurity strategy to protect against malicious threats while also allowing access to important data and systems in an efficient manner that is compliant with any applicable laws or regulations.

7. Stay organized using project planning software to ensure tasks are always being tracked accurately from beginning to end across different areas of the organization.

Things to Avoid in the IT Department

1. Do not neglect regular maintenance and updates on all systems, software, and hardware. Failing to keep up with these tasks can lead to security risks and problems down the road.

2. Do not overlook user training when it comes to introducing new systems or processes. Without proper knowledge and understanding of these tools, employees won't be able to use them correctly or efficiently.

3. Avoid making large changes quickly as this could create confusion and leave your team feeling overwhelmed. Make sure to think things through, test out ideas before implementation, and communicate any changes with everyone involved promptly.

4. Do not cut corners when it comes to cybersecurity measures such as installing firewalls, encrypting data, or regularly backing up important information. Doing so could make your organization vulnerable to attacks that can cause major damage.

5. Avoid siloing departments by establishing clear channels of communication between different teams within the IT department so that everyone is on the same page about goals and progress being made towards them.

6. Do not skimp on customer service when dealing with users who need assistance with technology-related issues such as software malfunctions or internet connectivity problems. Respond promptly and provide help in a friendly manner to ensure user satisfaction.

7. Do not forget about data privacy regulations like GDPR or HIPAA if you handle data related to customers, patients, or clients. Making sure that you adhere to all relevant laws will protect both you and

your organization from potential legal repercussions stemming from any violations.

8. Avoid overlooking long-term strategies when problem-solving instead of focusing too much on quick fixes for short-term gains. This could save your team time in the future by preventing further headaches from arising due to ill-conceived solutions that didn't address underlying issues effectively.

9. Do not leave routine tasks unattended for too long as this can lead to bigger problems down the line if not addressed promptly. Staying on top of maintenance such as patching systems, updating software, checking logs, etc. will help ensure things run smoothly while eliminating potential threats or errors before they become serious issues.

10. Do not forget about industry best practices even if they seem mundane at first glance because following them can often lead to safer usage habits which prevent accidents due to negligence or unfamiliarity with specific technologies used within an organization

Essential Skills of an IT Director

In a rapidly evolving technological world, an effective IT director or Chief Technology Officer (CTO) is crucial to the success of any organization. Charged with the responsibility of overseeing the entire IT infrastructure, an ideal IT director should harness innovation, anticipate future needs, and execute strategies to drive businesses to greatness. With endless possibilities, how do we identify the perfect IT director or CTO? This comprehensive guide will delve into the essential skills and attributes of an ideal IT director/CTO.

1. Technical Expertise:
An IT director/CTO must possess a deep understanding and knowledge of the technical aspects within the domain. They must be familiar with various programming languages, system architectures, databases, and network technologies, among others. Staying up to date on emerging trends and technologies is vital as they will be responsible for guiding the organization's technical roadmap.

2. Strategic Thinking:
An IT director/CTO must be able to see the bigger picture, anticipate future technology trends, and adapt the organization accordingly. This involves analyzing the company's current technological standing,

envisioning future possibilities, and developing a roadmap to reach these objectives.

3. Leadership:
Great leadership is an indispensable quality of an ideal IT director/CTO. They must inspire, motivate, and develop the IT team towards achieving short and long-term goals. They should cultivate a positive work environment that fosters creativity, encourages learning, and promotes talent growth and retention.

4. Project Management:
An IT director/CTO must be adept at managing multiple projects simultaneously. This involves setting realistic timelines, budgeting, resource allocation, and risk management. By efficiently managing projects, IT directors ensure that deadlines are met, and the outcome is aligned with the organization's needs and goals.

5. Communication Skills:
Effective communication is central to an IT director/CTO's success. It is their responsibility to articulate the technical needs and capabilities of the organization to both technical and non-technical stakeholders. They must be able to present complex information clearly, concisely, and persuasively to guide decision-making, secure budgets, and gain stakeholder buy-in.

6. Collaboration:
An ideal IT director/CTO should be capable of building strong collaborative relationships across various departments. They must work closely with the executive management team to align IT strategies with the company's overall business objectives. Also, they should foster cross-functional teams to enable knowledge sharing, drive innovation, and deliver results in a timely manner.

7. Business Acumen:
Understanding the intricacies of the organization's business is vital for an IT director/CTO. They should have substantial knowledge of the company's industry, target market, competitive landscape, and revenue streams. This business acumen enables IT directors/CTOs to make informed strategic decisions that facilitate the growth and profitability of the company.

8. Adaptability:
In a constantly evolving technological landscape, adaptability is an essential quality for an IT director/CTO. They should not only respond swiftly to changes within their field but also lead the organization through the adoption of new technologies in a seamless manner. Being open to change and fostering a culture of continuous improvement is crucial for the ongoing success of the organization.

9. Critical Thinking and Problem-solving Skills:
IT directors/CTOs must be skilled in identifying issues, analyzing complex situations, and developing effective solutions. They must be able to think critically, question assumptions, and remain open to diverse perspectives

The Most Important Asset is Critical Thinking

In the fast-paced and ever-evolving world of technology, an IT director must wear many hats and juggle numerous responsibilities. Among communication, adaptability, technical expertise, and project management, it may be tempting to place equal importance on all these skills. However, critical thinking and strategic thinking stand out as the most vital skills for an IT director to possess. These skills contribute to a more effective and significant impact on an organization's long-term success.

Critical thinking, refers to the ability to analyze situations and problems in a logical and unbiased manner, recognize patterns, and understand solutions to apply them to various scenarios. In the realm of information technology, critical thinking can prove invaluable for various reasons:

1. Collaborative Decision-making: IT directors work with cross-functional teams that often bring diverse perspectives and expertise to the table. Critical thinking enables an individual to understand different angles and engage in conversations while evaluating essential factors, risk assessments, and implications. This allows for more effective decision-making, with a clearer outcome and broader company impact.

2. Innovation: In an industry where innovation is key to staying ahead of the curve, critical thinking encourages individuals to question the status quo, identify gaps in processes, products, or services, and create new ways to improve them. IT directors that possess critical thinking skills are often more effective in driving change and staying ahead of their competitors.

3. Dealing with Complexity: IT directors need to handle complex situations and evaluate multiple variables at once. Critical thinking helps them break down problems into smaller, manageable components and approach them systematically. This not only reduces stress associated with complexity but also ensures that the most optimal solutions are delivered consistently.

For example, a critical thinker IT director may recognize a recurring issue in the company's workflow and decide to implement an automated solution. While this strategic decision might require an initial investment, it will greatly benefit the entire company in the long run, by streamlining operations and providing valuable insights into further optimization.

In conclusion: In a rapidly evolving IT landscape, having a critical thinker as an IT director can significantly contribute to the success of a company. By promoting collaborative decision-making, driving innovation, and navigating complexity, critical thinking empowers IT directors to stay agile and drive strategic decisions that benefit the entire organization.

Tips for Enhancing Critical Thinking Skills:

1. Challenge assumptions and seek diverse perspectives – regularly question your beliefs and consider alternative points of view.

2. Be open-minded and receptive to new ideas – encourage yourself and your team members to think outside of the box.

3. Always seek evidence and data to support decision making – validate assumptions with information to minimize risks.

4. Continuously refine and improve critical thinking abilities – pursue relevant training, courses, and certifications to stay up to date with best practices in critical and strategic thinking.

Unraveling the Inspiration and Aspiration Connection

The world of technology is an ever-changing landscape, with innovations and transformations occurring at breakneck speeds. In such an environment, IT professionals and departments are constantly challenged to stay ahead of the curve and adapt to disruptive changes. This brings to light the importance of inspiration and aspiration for IT departments. In this section, we will delve into the distinction between these two concepts and examine how best to synergize inspiration and aspiration to drive progress and success in the IT sphere.

Defining Inspiration and Aspiration

To begin, let us define the terms 'inspiration' and 'aspiration.' Inspiration refers to the mental stimulation an individual or group experiences when encountering something (an idea, event, or situation) that ignites their imagination, creativity, or passion. It acts as a catalyst, triggering a change in one's previously held beliefs, ideas or perceptions. In the context of the IT department, inspiration can be sourced from various channels such as industry breakthroughs, creative problem-solving, or innovative collaborations.

On the other hand, aspirations are the goals, dreams, or ambitions that an individual or organization wishes to achieve in the future. It is a clear vision of what they aim to accomplish and involves a roadmap for moving from their current state toward these desired outcomes. When it comes to IT departments, aspirations can be both tangible, like creating new software or breaking into new markets, or intangible, like fostering a culture of innovative thinking or nurturing employee growth.

Harnessing Inspirations to Fuel Aspirations

Understanding the difference between inspiration and aspiration is the first step in harnessing their potential to propel IT departments to greater heights. Here are a few examples of leveraging inspirations for cultivating aspirations:

1. **Tech Conferences:** Attending industry events can inspire IT professionals through exposure to cutting-

edge technology, emerging trends, and conversations with thought leaders. This inspiration can lead to the formation of new aspirations like adopting new technologies, developing innovative products, or expanding skill sets.

2. **Analyzing Competitors:** Studying the strategic moves and successes of industry competitors can inspire IT departments to reassess their current approach and aspire to elevate their offerings, streamline processes or implement more efficient systems.

3. **Collaboration and Cross-Pollination:** Interacting and collaborating with professionals from other disciplines or industries can inspire IT departments with new perspectives, best practices or creative solutions that they can aspire to integrate into their operations.

Best Practices for Creating Aspiration-Driven Inspirations

To conclude, here are some good practices that IT departments can employ to effectively utilize inspiration to drive their aspirations:

1. **Encourage Exploration:** Establish an organizational culture that promotes learning, experimentation and risk-taking to allow for the pursuit of inspirations that can lead to aspirational outcomes.

2. **Create a Vision:** Clearly articulate your IT department's aspirations and ensure all team members are aligned towards these goals.

3. **Celebrate Successes:** Recognize and celebrate both inspirational and aspirational achievements within the department to cultivate a culture of continuous growth and improvement.

4. **Foster Open Communication:** Encourage the free flow of ideas, questions, and inspirations, keeping an open mind to new possibilities that can contribute to the department's aspirations.

5. **Be Adaptable:** Embrace change and remain flexible, utilizing new inspirations to reevaluate and refine aspirations as necessary.

Understanding the interdependence of inspirations and aspirations, permits IT directors & managers to have at their disposal powerful tools which can create successful outcomes.

Manage the People by Example

This is a powerful sentiment that many may avoid comprehending at different points in life. However, no matter how intensively one evaluates the strategies of proficient management, nothing will be as prosperous as when led by an exemplary manager. This not only applies to all managerial roles but also to those who direct IT departments or other businesses - they are no exception to this regulation.

How Motivate the Team

How to Keep Your Team Motivated Amid Challenges and Routine Tasks

In today's fast-paced and competitive environment, keeping your team motivated is more important than ever. When faced with challenges such as reluctance to work routinely, reluctance to answer phones, and poor customer service interactions, it is essential to maintain a positive and energetic atmosphere in the workplace. Drawing on personal experience and best practices, here are some tips on how to keep your team inspired and excited about their job processes:

1. Cultivate a supportive and empowering work environment: Encourage open communication within the team, recognize individual and group achievements, and provide opportunities for skill development. When a team feels valued and supported, they are more likely to work together to overcome challenges and deliver excellent customer service.

2. Set clear expectations and goals: Make sure each team member understands their responsibilities and the impact of their work on the organization's goals. Provide constructive feedback regularly and help your team members set achievable and measurable objectives to prevent burnout and maintain motivation.

3. Encourage creativity and problem-solving: Providing your team with custom software or tools that enable them to be more creative and efficient in their job processes can reinvigorate their enthusiasm. Facilitate brainstorming sessions to identify innovative solutions that can streamline workflows, reduce monotonous tasks, and improve customer interactions.

4. Celebrate progress and milestones: Regularly acknowledge the hard work of your team, and celebrate their accomplishments, both large and small. Organize team social events or team-building activities to foster camaraderie and make work feel more enjoyable.

5. Foster a sense of ownership: Empower your team members by involving them in decision-making processes and giving them a voice in their work environment. When employees feel a sense of ownership over their tasks and projects, they are more likely to stay committed and motivated.

6. Keep the lines of communication open: Ensure that your team members receive the necessary information and updates concerning their tasks and the customers they are dealing with. When employees are well-informed, they are better equipped to handle customers' issues and provide exceptional service experiences.

7. Implement a reward and recognition system: Show team members that you appreciate their hard work and dedication through incentives, pay raises, or other recognition initiatives. Reinforce positive behavior by acknowledging a job well done, fostering a sense of pride and accomplishment within the team.

8. Encourage continuous learning and growth: Provide opportunities for your team members to expand their skill sets, engage in training seminars, and take part in workshops to stay up to date with the latest industry trends. When employees feel challenged and encouraged to develop professionally, motivation will soar.

By implementing these best practices, you can keep your team motivated and energized in the face of challenges and repetitive tasks. A motivated team is vital to maintaining excellent customer service, positive interactions, and achieving success in any organization.

People Management

Many companies are confronted with the challenge of managing their people. People management is perhaps one of the most difficult tasks a company can face, as it involves taking care of not only employees' professional development but also their personal life. It requires both a deep understanding of human behavior and an expertise in managing complex situations.

It is essential to recognize that your staff needs the same level of attention as your customers. They do possess influence in requesting and receiving it, so it's important to be attentive. Avoid unintentionally making a top employee turn average or even worse - disappointed; pay close attention to their requirements and address any pending issues immediately upon arising.

It is crucial to explain why your actions are taken, even if the conversation may be uncomfortable. Always ensure that evaluations are given promptly and remain impartial throughout; it is necessary for everyone involved to understand that all comments provided are fair and consistent.

Things to Remember

Lead by Example

Leadership is a critical aspect of any organization, community, or group. A leader's actions, behavior, and values have a profound impact on those around them. They set the tone for the entire group and play a crucial role in determining its success.

One of the most important traits of a great leader is the ability to lead by example. This means that leaders must model the kind of behavior and values they want to see in their followers. Whether it is ethics, work ethic, or behavior, a leader cannot expect others to follow them if they do not set the proper example themselves.

For example, in a workplace setting, leaders who arrive at work early, demonstrate a strong work ethic, and put in the extra effort necessary to get the job done often inspire their direct reports to do the same. In this way, they create a culture of hard work and productivity that benefits everyone.

Similarly, a leader who values transparency and honesty will be respected by their followers and inspire them to be truthful and open as well. Conversely, if a leader is dishonest or tries to hide information, this behavior will trickle down and create a workplace where dishonesty and mistrust are rampant.

Furthermore, leaders who lead by example demonstrate to their followers that they are willing to do what it takes to get the job done. They are not above performing menial tasks or rolling up their sleeves when necessary. This helps to create a sense of unity and teamwork, as everyone feels like they are part of a common goal.

Leading by example is a critical trait of any great leader. By setting the proper example, leaders can inspire their followers to adopt the same behavior and values, creating a culture of excellence and success. Leaders who fail to lead by example risk creating a culture where bad behavior, dishonesty, or lack of accountability are the norms. Therefore, leaders must always strive to set the best possible example for their followers to follow.

Motivate and Inspire Your People

Motivated and inspired individuals are more engaged in their work and are more likely to go above and beyond their job duties to meet goals and target achievements. This is because they see that what they are doing has a higher purpose and a greater impact. Furthermore, individuals who are inspired and motivated feel empowered and confident about their roles, which helps to foster a positive work environment.

As a leader, setting a leadership example is the first step in inspiring and motivating your team. A leader who consistently demonstrates a strong work ethic, passionate commitment to the team's vision and mission, and a positive attitude can inspire trust and loyalty in their followers. Leaders who lead by example can earn respect and build credibility with their team, which helps to motivate and inspire their followers.

Here are some practical examples of setting a leadership example:
Show empathy. As a leader, it is essential to show empathy to members of the team. Listening to their

concerns and addressing their issues helps to create a positive work environment where everyone feels heard and appreciated.

Be a team player. Leaders who are willing to participate equally in teams help to inspire and motivate their team members. It shows that they are not above performing menial tasks or helping others when needed.

Encourage learning and development. Leaders who encourage their team members to learn new skills and continuously develop themselves are more likely to have a team that is motivated and inspired.

Ideas to inspire and motivate your team:

- Providing regular and effective feedback to your team helps to boost their confidence and provides them with a sense of direction. Feedback provided positively and constructively builds a positive work environment.

- Celebrate successes. Recognize the team's achievements, no matter how small, and celebrate them. Celebrating successes helps build a positive work environment, motivates team members, and reinforces the value of their hard work.

- Providing opportunities for your team members to learn and advance their skills shows that you are invested in their growth and development. This helps to inspire and motivate the team to learn new things and stay engaged.

Motivating and inspiring your team is a vital aspect of leadership. While policies and standards are essential, they should not be the only way you manage your team. Rather, focus on setting an inspiring leadership example and finding ways to motivate and inspire your team. By doing this, you can build a positive work environment, motivate your team to go above and beyond, and ultimately achieve success as a team.

Failures are the Ladder to Success

Failures, often considered as a negative aspect, could be the ladder to success. The phrase "fail forward" refers to the idea that when we fall or fail, we are falling and failing to learn and grow. Falling is just another step in the journey to build something incredible. Failure helps us to understand our weaknesses, learn from mistakes and allows us to become stronger and more resilient.

The iconic story of Thomas Edison is a testament to the potency of perseverance through adversity. He tried more than 10,000 separate attempts before he invented the light bulb – but never gave up hope! Despite being unsuccessful in his numerous experiments, each 'mistake' was an essential stride towards achieving his ambition: something that would revolutionize our world for good. Eventually, this unwavering commitment and resilience paid off with the emergence of incandescent lightbulbs - altering civilization's future forevermore.

Another example can be the well-known brand, Airbnb. The founders of Airbnb, Brian Chesky, and Joe Gebbia have shared their stories of how they struggled to get enough users in the early days of the company. They even had to create fake listings to make it seem like there were more available rooms on the site. However, the founders took this failure as an opportunity to learn more about their users and their needs. This helped them to create a personalized experience that made people feel more connected to the brand. The result of their determination was a revolutionary platform that has completely changed the hospitality industry.

In terms of business, it is essential to learn from failures and grow from it. By shifting focus towards the lessons learned, both as an individual and as a company, you can quickly find ways to adapt during challenging times. Competitors, technology, or market trends do not definitively determine success or failure; instead, it is the constant improvement after each discovered failure that sets the foundation.

A failed concept doesn't necessarily mean that the vision or the goal was unachievable; rather, it indicates that the process, approach, or execution was what ultimately led to the failure. Therefore, refining the strategy or approach, analyzing the market, and understanding consumer preferences are crucial. Adopting a more practical approach to the next move based on learned lessons would always lead to success eventually.

In conclusion, failures are not the end of a journey, instead, they are an opportunity to learn and grow. Successful people understand this and have learned to look at failures as lessons that will teach and help build resilience. The ability to learn from failure and embrace change is one of the hallmarks of successful individuals and entrepreneurs. Therefore, it is essential to embrace failures, analyze them, and use the insights gained to iterate on our next steps.

Be a Servant Leader

Servant leadership is a management philosophy that prioritizes the success and well-being of individuals rather than achieving organizational goals at all costs. It is a leadership style that resonates with the values of empathy, humility, and ethical behavior. A servant leader, as the name suggests, is someone who leads by serving their team or organization, putting others' needs before their own.

The concept of servant leadership was first proposed by Robert K. Greenleaf in 1970, who believed that the most effective leaders are those who serve their followers rather than the other way around. Servant leaders recognize that their role is not to dominate, control or micromanage, but rather to empower and enable others to achieve their full potential.

One of the key attributes of a servant leader is empathy, which involves putting oneself in the shoes of others to understand their needs, concerns, and motivations. Servant leader listens to their team, learns from them, and responds to their needs by providing guidance, support, and resources to help them succeed. This creates a culture of collaboration, respect, and trust, which is essential for achieving long-term success.

Another important characteristic of a servant leader is humility, which involves admitting mistakes, acknowledging the contributions of others, and putting the greater good above personal ambition. This fosters an environment in which everyone feels valued, respected, and appreciated, regardless of their position or status within the organization.

In addition to empathy and humility, servant leaders practice ethical and responsible decision-making, using their positions of power to promote the greater good rather than their interests. They prioritize social responsibility, ethical conduct, and sustainability, emphasizing that profitability should not come at the expense of social or environmental impact.

Being a servant leader is important because it creates a positive work culture that fosters growth, innovation, and collaboration. It helps build trust and loyalty among team members, maximizing their potential and productivity. It also contributes to the overall success of an organization by promoting a shared sense of purpose and vision.

In today's rapidly evolving business environment, the value of servant leadership cannot be overstated. As organizations grapple with the challenges of a rapidly changing market and a diverse workforce, leaders who prioritize the needs of their team rather than their own will be better equipped to navigate the complexities of the business world. Thus, being a servant leader is not only morally and ethically right, but it is also a sound business strategy that leads to long-term success.

Biographies

Dr. Julia Diakakis

Dr. Julia Diakakis, a highly respected and decorated educator and businesswoman, boasts an impressive career spanning over 35 years. Born into a military family in Bremerhaven, Germany, Dr. Diakakis was raised throughout Europe providing her a diverse upbringing which laid the foundation for a versatile professional journey.

A proud graduate of Ohio University, she holds a B.S. in Education with a minor in English. Her certifications span various subjects including English, Math, History, Social Science, Earth Space Science, and Elementary Education with endorsements in Reading and English Speakers of Other Languages (ESOL). Her passion for education led her to different roles in educational institutions ranging from elementary to high school levels, earning her recognition as one of Florida's most esteemed educators. Furthering her passion for education, she has served on dissertation committees as the content expert for doctoral students.

Parallel to her educational career, Dr. Diakakis has been maintaining an IT business since 1994 and served the Brevard School Board as an IT Network Administrator for seven years. This combination of education and technology expertise uniquely positions her at the intersection of these two critical fields.

Dr. Diakakis further expanded her skill set by obtaining a master's degree in educational administration, leading to her principal certification and an administrative role for about 15 years. She didn't stop there; her boundless passion for education, business, and information technology propelled her to pursue a Doctorate in Organizational Leadership and Development.

A member of prestigious societies such as Alpha Chi Honor Society, Kappa Delta Pi Honor Society, and

National Society of School Administrators, Dr. Diakakis has made significant contributions to developing computer lab programs for Brevard Public Schools in Florida. She also pioneered the first fully operational Discourse Study Computer classroom in Florida. Her extensive work with numerous private companies throughout her career testifies to her business acumen.

Dr. Diakakis' unique blend of educational knowledge and business insights makes her a formidable force in her field. Her life and career serve as an inspiration for many who aspire to make a difference in the realms of education and business.

Saki Diakakis

Born in Athens, Greece, Saki Diakakis embarked on his journey with a classic high school education followed by military service as an Amphibian Green Beret. In 1989, Saki immigrated to the United States to nurture his passion for aviation and began training as a private pilot, initially aiming for a career as a commercial pilot.

However, life had other plans, and he found myself drawn towards the burgeoning field of computers, networks, and software development. Saki returned to academia, this time immersing himself in computer science, and subsequently ventured into diverse IT industries.

From the early 1990s, his career path led him to work with local city government bodies as a Network Systems Analyst and Software Developer. His professional portfolio expanded across multiple sectors including county, city, and state governments, in areas of advertising, space subcontractors, home builders, and water treatment plants, among others. During his time with city government he designed and built the TV station for the City of Cocoa Beach, marking a significant milestone in his career.

During a nine-year stint in the advertising industry, Saki managed central networks and developed integral systems for the national yellow page company. As a consultant, he contributed to data analysis and application development. Over the years, he has developed an array of commercial applications encompassing document management systems, CRM, billing, building department permits, fleet management, and more. Adding another feather to the cap, he pioneered one of the first surveillance wireless undercover applications in the United States.

For the past 12 years, Saki has been associated with a local Managed Services Provider, specializing in computer networks, telephony, and cybersecurity applications. His services have benefited hundreds of businesses on local, federal, and international levels.

Saki's extensive formal education and practical experience have equipped him with knowledge that many businesses and stakeholders consider invaluable for their success. His is committed to leveraging this expertise to continue the contribution to the field of technology and businesses

NOTES:

NOTES:

Made in United States
Orlando, FL
04 November 2026

90201201R00092